ESSENTIAL SKILLS FOR GCSE

Combined Science

Dan Foulder
Nora Henry
Roy White

Hachette UK's policy is to use papers that are natural, renewable and recyclable products and made from wood grown in sustainable forests. The logging and manufacturing processes are expected to conform to the environmental regulations of the country of origin.

Orders: please contact Bookpoint Ltd, 130 Park Drive, Milton Park, Abingdon, Oxon OX14 4SE. Telephone: (44) 01235 827720. Fax: (44) 01235 400454. Email education@bookpoint.co.uk Lines are open from 9 a.m. to 5 p.m., Monday to Saturday, with a 24-hour message answering service. You can also order through our website: www.hoddereducation.co.uk

First published in 2019 by

Hodder Education,

An Hachette UK Company

Carmelite House

50 Victoria Embankment

London EC4Y 0DZ

www.hoddereducation.co.uk

Impression number 10 9 8 7 6 5 4 3 2 1

Year 2023 2022 2021 2020 2019

Typeset by Integra Software Services Pvt. Ltd., Pondicherry, India

Printed by Replika Press Pvt. Ltd., Haryana, India

A catalogue record for this title is available from the British Library.

ISBN: 978 1 510 45999 1

FSC www.fsc.org MIX Paper from responsible sources FSC™ C104740

Contents

How to use this book

Welcome to *Essential Skills for GCSE Combined Science*. This book covers the major UK exam boards for Science: AQA, Edexcel (including Edexcel International GCSE), OCR 21st Century and Gateway, WJEC/Eduqas and CCEA. Where exam board requirements differ, these specifics are flagged. This book is designed to help you go beyond the subject-specific knowledge and develop the underlying essential skills needed to do well in GCSE Science. These skills include Maths, Literacy, and Working Scientifically, which now in recent years have an increased focus.

- The Maths chapter covers the five key areas required by the government, with different Science-specific contexts. In your Science exams, questions testing Maths skills make up to 20% of the marks available with a 1:2:3 ratio for Biology, Chemistry and Physics.
- The Literacy chapter will help you learn how to answer extended response questions. You will be expected to answer at least one of these per paper, depending on your specification and they are usually worth six marks.
- The chapter on Working Scientifically covers the four key areas that are required in all GCSE sciences.
- The Revision chapter explains how to improve the efficiency of your revision using retrieval practice techniques.
- Finally, the Exam Tips chapter explains ways of improving your performance in the actual exam.

To help you practise your skills, there are three exam-style papers at the end of the book, with another three available online at **www.hoddereducation.co.uk/EssentialSkillsCombinedScience**. While they are not designed to be accurate representations of any particular specification or exam paper, they are made up of exam-style questions and will require you to put your maths, literacy and practical skills into action.

Key Features

In addition to Key term and **Tip** boxes throughout the book, there are several other features designed to help you develop your skills.

A Worked example

These boxes contain questions where the working required to reach the answer has been shown.

A Expert commentary

These sample extended responses are provided with expert commentary, a mark and an explanation of why it was awarded.

B Guided questions

These guide you in the right direction, so you can work towards solving the question yourself.

B Peer assessment

These activities ask you to use a mark scheme to assess the sample answer and justify your score.

C Practice questions

These exam-style questions will test your understanding of the subject.

C Improve the answer

These activities ask you to rewrite a sample answer to improve it and gain full marks.

Answers to all questions can be found at the back of the book. These are fully worked solutions with step-by-step calculations included. Answers for the three online exam-style papers can also be found online at **www.hoddereducation.co.uk/EssentialSkillsCombinedScience**.

★ **Flags like this one will inform you of any specific exam board requirements.**

1 Maths

Maths is an important part of GCSE Science, and specific marks are now awarded in the exams for how well you answer maths-based questions. The important thing to remember is that these are all skills you will be using in GCSE Maths – just in a different context, namely science.

For example, common maths questions in science might ask you to draw graphs, calculate means and work out probabilities (e.g. from genetic crosses). As you can see, none of these skills should be alien to you.

This chapter aims to take you through all the maths skills needed in order to successfully complete your GCSE Science course.

» Units and abbreviations

Scientific quantities are measured using units. Units are very important in science. Without them, numerical values are often meaningless, and leaving them out will cost you marks in the exam. You should ensure that you use appropriate units across all calculations and data handling.

Laboratory work is central to GCSE Science. Throughout your course there will be many opportunities for you to use practical work to investigate, record and process data. Some measurements recorded in experimental work may be qualitative and would not include a numerical value.

A range of quantitative units are used in GCSE Science. Scientists use the SI system of measurement. This system is based on the seven fundamental base units shown in Table 1.1. Whenever possible, you should use the internationally recognised units. You should familiarise yourself with both the units and their abbreviations.

Table 1.1 Base units in GCSE Science

Measurement	Unit	Abbreviation
mass	kilogram	kg
length	metre	m
time	second	s
current	ampere (amp)	A
temperature	degree Celsius	°C
amount of substance	mole	mol
luminous intensity	candela	cd

All other SI units are combinations of the base units. These combinations are called derived units (see Table 1.2).

Compound units are those composed of more than one unit. For example:

- Concentration is measured in mol/dm^3 (mol per dm^3)
- Energy change is measured in kJ/mol (kilojoule per mol)

Key terms

Qualitative: Descriptions of how something appears, rather than with figures or numbers.

Quantitative: Measurements such as mass, temperature and volume, involve a numerical value. For these quantitative measurements, it is essential that units are included because stating that the mass of a solid is 0.4 says very little about the actual mass of the solid – it could be 0.4 g or 0.4 kg.

Base units: The units on which the SI system is based.

Derived units: Combinations of base units such as m/s and kg/m^3.

Tip

SI stands for Système International. You don't need to know about the history of the system – you just need to know the units.

Table 1.2 GCSE derived units

Physical quantity	Derived unit	Abbreviation
area	square metres	m^2
volume	cubic metres	m^3
density	kilogram per cubic metre	kg/m^3
pressure	pascal	Pa
specific heat capacity	joule per kilogram per degree Celsius	J/kg°C
specific latent heat	joule per kilogram	J/kg
speed	metre per second	m/s
force	newton	N
gravitational field strength	newton per kilogram	N/kg
acceleration	metre per squared second	m/s^2
frequency	hertz	Hz
energy	joule	J
power	watt	W
electric charge	coulomb	C
electric potential difference	volt	V
electric resistance	ohm	Ω
magnetic flux density	tesla	T

It would be inappropriate to give, say, the mass of a postcard in kilograms, so scientists often use smaller (submultiple) and larger (multiple) units in calculations. The common ones are shown in Table 1.3.

Table 1.3 Submultiple and multiple units in GCSE Science

Prefix name	Symbol	Meaning/Standard form	decimal
tetra	T	$\times 10^{12}$	1 000 000 000 000
giga	G	$\times 10^{9}$	1 000 000 000
mega	M	$\times 10^{6}$	1 000 000
kilo	k	$\times 10^{3}$	1000
centi	c	$\times 10^{-2}$	0.01
milli	m	$\times 10^{-3}$	0.001
micro	μ	$\times 10^{-6}$	0.000 001
nano	n	$\times 10^{-9}$	0.000 000 001

These examples show sensible units to measure the objects:

- A postcard measures 15 cm by 8 cm
- The distance between London and Birmingham is about 190 km
- The diameter of a one pound coin is 22.5 mm

Converting between units

Addition and subtraction of values can only be done if they are expressed in the same units. For example, the mass of an evaporating basin (24 g) cannot be added to the mass of copper oxide (3000 mg) to give a total mass, as the units are different. If the units are different they must be converted to a common unit before being added together. In this example, the mass of copper oxide must first be converted from 3000 mg to 3 g and then added to the mass of the evaporating basin in grams (24 g) to give a total mass of 27 g.

You will need to be able to convert between different volume and mass units as outlined below.

Key terms

Submultiples: Fractions of a base unit or derived unit, such as centi- in centimetre.

Multiples: Large numbers of base or derived units, such as kilo- in kilogram.

Tip

For more on standard form see page 8.

Tip

There are also a few physical quantities that are simple ratios and have no unit.

At GCSE these are:

- efficiency
- magnification
- transformer turns-ratio

Remember you do *not* need to put a unit symbol with these quantities.

Tip

Above 9999, groups of digits are usually separated in groups of three by a space; for example, 10 000. You should avoid using a comma to separate groups of digits because, outside the UK and the USA, a comma is often used as a decimal point. Mistaking a comma as the position of a decimal point when administering drugs or designing a bridge could be disastrous.

Volume

Volume is usually measured in centimetres cubed (cm^3), decimetres cubed (dm^3) or metres cubed (m^3).

$$1000\ cm^3 = 1\ dm^3$$

$$1000\ dm^3 = 1\ m^3$$

You need to be able to convert between volume units, particularly for calculations on solution volume and concentration. The flow scheme in Figure 1.1 will help you to convert between volume units.

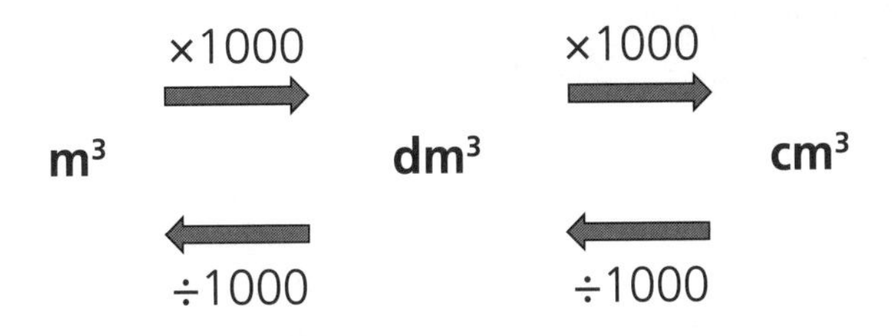

▲ Figure 1.1 Converting between volume units

Mass

Mass can be measured in milligrams (mg), grams (g), kilograms (kg) and tonnes (t).

1 tonne = 1000 kg

1 kilogram = 1000 g

1 gram = 1000 mg

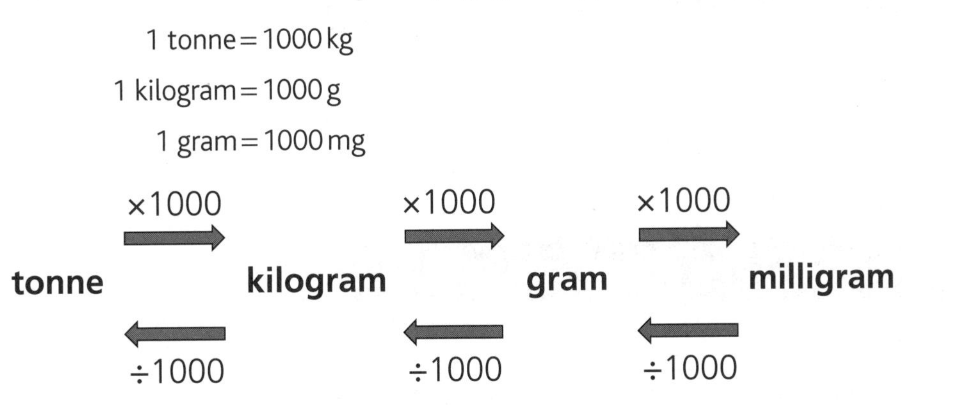

▲ Figure 1.2 Converting between mass units

> **Tip**
>
> Use common sense when converting between units. A kilogram is bigger than a gram so when converting from kilograms to grams you would expect to get a larger number of grams after your calculation.

A Worked examples

1 a **Convert 35 cm³ to dm³.**

Look at Figure 1.1. To convert from cm^3 to dm^3 divide by 1000.

$$\frac{35}{1000} = 0.035\ dm^3$$

b **Convert 1.5 dm³ to cm³.**

Look at Figure 1.1. To convert from dm^3 to cm^3 multiply by 1000.

$1.5 \times 1000 = 1500\ cm^3$

c **Convert 325 mg to grams.**

Look at Figure 1.2. To convert from mg to g divide by 1000.

$$\frac{325}{1000} = 0.325\ g$$

d **Convert 4.3 kg to grams.**

Look at Figure 1.2. To convert from kg to g multiply by 1000.

$4.3 \times 1000 = 4300\ g$

e Convert 2.2 tonnes to grams.

There are two conversions needed: tonne → kilogram → gram

Step 1: To convert from tonnes to kilograms multiply by 1000.

$2.2 \times 1000 = 2200$ kg

Step 2: To convert from kilograms to grams multiply by 1000.

$2200 \times 1000 = 2\,200\,000$ g

Tip

Refer to Figure 1.2 to help with this conversion.

B Guided questions

1 **Convert 1.2 dm^3 to cm^3.**

To convert from dm^3 to cm^3 multiply by 1000.

$1.2 \times 1000 =$ .... cm^3

2 **Convert 8.2 tonnes to g.**

Two conversions are needed here:

tonnes → kg → g

Step 1 To convert from tonnes to kg, multiply by 1000.

$8.2 \times 1000 =$ .... kg

Step 2 Then, to convert from kg to g, multiply by 1000.

.... $\times 1000 =$ .... g

Tip

Always think about your answer; a dm^3 is a larger unit than a cm^3 so you would expect to get a larger number of cm^3 when converting that way.

C Practice question

3 Carry out the following unit conversions.

a 1.2 dm^3 to cm^3

b 420 cm^3 to dm^3

c 3452 cm^3 to dm^3

d 4.4 t to g

e 4 kg to g

f 3512 g to kg

Calculations that often involve conversion of units

There are several topic areas in which you may often need to convert units. In Chemistry, for example, you may have to deal with conversions when looking at:

- reacting volumes of gases, where you may need to calculate moles of gas using the equation:

$$\text{amount (in moles)} = \frac{\text{volume }(dm^3)}{24}$$

- amounts of substances, where you may need to calculate moles using the equations:

$$\text{amount (in moles)} = \frac{\text{mass (g)}}{A_r} \quad \text{or} \quad \text{amount (in moles)} = \frac{\text{mass (g)}}{M_r}$$

Tip

A_r is the relative atomic mass and is found on the Periodic Table for each element. M_r is the **relative formula mass**. These values do not have units.

A Worked example

Calculate the amount, in moles, present in 2.4 tonnes of magnesium.

To calculate the amount in mole use the following equation:

$$\text{amount (in moles)} = \frac{\text{mass (g)}}{A_r}$$

Before using this expression, the mass of magnesium must be converted from tonnes to grams.

tonnes → kilograms → grams

×1000 ×1000

Step 1 mass of magnesium in grams = 2.4×1000×1000=2 400 000 g

Step 2 $\text{amount (in moles)} = \frac{\text{mass (g)}}{M_r}$

$$= \frac{2400000}{24} = 100\,000\,\text{mol}$$

Key term

Relative formula mass, M_r: The sum of the relative atomic masses (A_r) of all the atoms shown in the formula.

Tip

Some examination boards may ask you to calculate the number of moles rather than the amount in moles – this is answered in the same way.

B Guided questions

1 **Calculate the amount in moles present in 9.8 kg of sulfuric acid, H_2SO_4, which has relative formula mass (M_r) 98.**

To calculate the amount in moles use the expression

$$\text{amount (in moles)} = \frac{\text{mass (g)}}{M_r}$$

Step 1 Convert the mass from kg to g by multiplying by 1000.

9.8×1000 = ...

Step 2 Substitute the mass in grams and the M_r into the equation to calculate your final answer.

$$\text{amount (in moles)} = \frac{\text{mass (g)}}{98} = \text{....................} = \text{....................}$$

2 **Calculate the amount in moles present in 48 000 cm^3 of nitrogen gas.**

To calculate the amount in moles of a gas use the expression

$$\text{amount (in moles)} = \frac{\text{volume (cm}^3\text{)}}{24}$$

Step 1 Convert the gas volume from cm^3 to dm^3 by dividing by 1000

$$\frac{48000}{1000} = \text{....................}\,dm^3$$

Step 2 Substitute the volume into the equation and calculate your final answer.

$$\text{amount (in moles)} = \frac{\text{volume (dm}^3\text{)}}{24} = \text{...............} = \text{...............}$$

Tip

This equation can only be used to calculate the moles of a gas if the volume is given.

C Practice questions

3 Calculate the amount in moles present in 6 kg of calcium (Ca).

4 Calculate the amount in moles present in 3.2 tonnes of calcium carbonate. Calcium carbonate has relative formula mass (M_r) = 100.

5 Calculate the amount in moles present in 17 kg of ammonia (NH_3).

6 Calculate the amount in moles present in 2.1 tonnes of iron(III) oxide (Fe_2O_3).

7 Calculate the amount in moles present in 0.592 kg of magnesium nitrate ($Mg(NO_3)_2$).

8 Calculate the amount in moles present in 7200 cm^3 of sulfur trioxide gas (SO_3).

» Arithmetical and numerical computation

Expressions in decimal form

When adding or subtracting data, decimal places are often used to indicate the precision of the answer. The term 'decimal place' refers to the numbers after the decimal point. The number of decimal places is the number of digits after the decimal point.

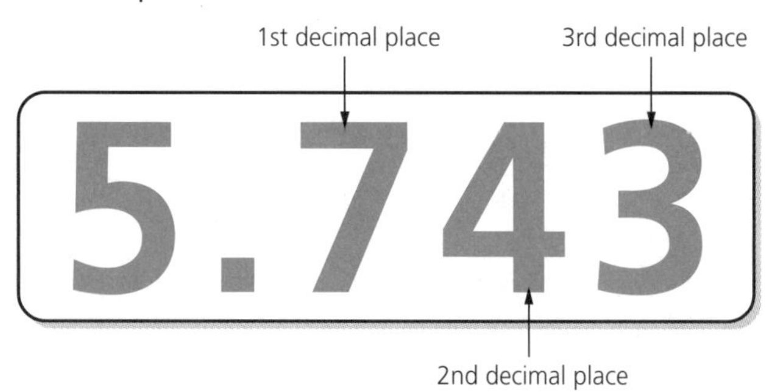

▲ Figure 1.3 Decimal places

The number 5.743 has three decimal places, while the number 10 has no decimal places.

Sometimes you are asked to present your answer to one or two decimal places. To do this you need to round your answer. For example:

- Rounding a number to one decimal place means there is only one digit after the decimal point.
- Rounding a number to two decimal places means there are two digits after the decimal point.

To round a number to a given number of decimal places, look at the digit after the last one you need (the decider figure) and

- if the next number is *5 or more*, round up
- if the next number is *4 or less*, do not round up.

For example, if you are rounding a number to two decimal places, it is useful to underline all numbers up to two numbers after the decimal point. This then focusses your attention on the next number, which helps with rounding.

Tip

Decimal numbers can also often be used to express fractions, for example $\frac{1}{2} = 0.5$.

Key terms

Decimal places: The number of integers given after a decimal point.

Integers: These are whole numbers, which includes zeros.

Decider figure: The integer after the number of decimal places required, it *decides* whether we must round up or not.

You should ensure that you round using the appropriate number of decimal places (d.p.) in answers to exam questions. This will probably depend on the number of decimal places used in the question or on the data provided.

In practical questions involving measurements, your answers should not have more decimal places than the least accurate measurement. For example, if a ruler is used to measure the area of the sides of a cube to the nearest 0.1 cm, the surface area (cm^2) and volume (cm^3) calculated using this value should not be given to more than one decimal place.

While most values have an exact number of decimal places, others might have recurring decimals (for example $\frac{1}{3} = 0.333333333$ recurring) or an infinite number of decimal places (like pi – π). These unusual values should always be rounded to the appropriate number of decimal places.

Key term

Recurring: When a number goes on forever.

An example of a recurring decimal would be if you divide the number 2 by the number 3 on your calculator. With this calculation, you may see 0.6666666666 on your display although more modern calculators display it as $0.\dot{6}$; the dot above the 6 shows that the 6 repeats forever. In either case, the rule for rounding is the same. For 3 d.p., we would write 0.6666666666 or $0.\dot{6}$ as 0.667. Note that, if there are two dots, then the numbers *between* the dots are repeated. So, $0.\dot{6}5\dot{2}$ means 0.652652652... This would be 0.7 to 1 d.p., 0.65 to 2 d.p. and 0.653 to 3 d.p.

A Worked example

A physicist measures the diameter of a metal rod as 0.7 cm to 1 d.p.

Giving both answers to 2 d.p., what is:

a **the smallest diameter that the rod could have had?**

Step 1 The smallest number would still have to be a value that the physicist rounded up, so the first decimal place would be a 6.

Step 2 The number following the 6 would have to be as small as possible, but still a number that allows us to round up.

Step 3 So, the smallest diameter is 0.65.

b **the largest diameter that the rod could have had?**

Step 1 The largest number must be small enough to not have been rounded up, so the first decimal place would be a 7.

Step 2 The number following the 7 would have to be as large as possible, but still a number that won't allow us to round up.

Step 3 So, the largest diameter is 0.74.

B Guided questions

1 **A root had a diameter of 0.345 cm. Write this diameter to one decimal place.**

The second decimal place is 4, so you should round down.

Root diameter to one decimal place = ...

2 **A toy car rolls down a slope and travels 20 cm in 6.4 seconds. Calculate the speed of the car in cm/s to 2 d.p.**

Step 1 Find the speed: $\frac{20\text{ cm}}{6.4\text{ s}}$ = ... cm/s

Step 2 Give the answer to 2 d.p: speed of car = ... cm/s

Tip

If creating results tables in your exam or a required practical, remember that all of the values should have the same number of decimal places.

C Practice question

3 Copper(II) sulfate solution was electrolysed for five minutes using copper electrodes. The table shows the mass of the copper anode and cathode before and after the electrolysis.

	anode	cathode
Mass of electrode before electrolysis/g	1.66	1.58
Mass of electrode after electrolysis/g	1.15	1.87

Calculate the mass of copper deposited to one decimal place.

4 A student weighs 630 N and the total area of his feet in contact with the ground is 205 cm^2.

Calculate the pressure he exerts on the ground, giving your answer in N/cm^2 to 1 d.p.

Tip

In this electrolysis experiment, copper is deposited on one electrode, which gets heavier, and the other electrode gets lighter.

Standard form

Scientists sometimes deal with very large or very small numbers. For example:

- the number of water molecules in a tablespoon of water is about 602 000 000 000 000 000 000 000 (an incredibly large number)
- the wavelength of an X-ray is about 0.000 000 000 1 metres (an incredibly small number)
- the energy that a plant receives from the Sun could be 18 00 000 kJ/m^2/yr
- a bacterial cell can have a diameter of 0.005 mm

Standard form is used to express very large or very small numbers so that they are more easily understood and managed. It's easier to say that a speck of dust weighs 1.2×10^{-6} grams than to say it weighs 0.000 001 2 grams or that a carbon-to-carbon bond has length 1.3×10^{-10} m than to say it is 0.000 000 000 13 m.

Standard form must always look like Figure 1.4:

'A' must always be between 1 and 10

'n' is the number of places the decimal point moves

$$A \times 10^n$$

▲ Figure 1.4 Standard form

'*n*' can also be thought of as the power of ten that *A* is multiplied by to equal the original number. It is important that you can convert standard form back to ordinary form and from ordinary form to standard form. See Table 1.3 on page 2 to see how prefixes are used to represent standard form.

Powers of 10

Powers of 10 give us a way to write these very large and very small numbers in a sort of short-hand format. For example, if we look at the calculation $10 \times 10 = 100$, we can see that two tens are multiplied together. We can, therefore, write the value of 100 as 10^2 or 1.0×10^2.

In the calculation $10 \times 10 \times 10 \times 10 = 10000$, we can see that four tens are multiplied together. We can write the value of 10000 as 10^4, or 1.0×10^4. The number of zeros translates into a power of 10 when each number is written in standard form. Powers are written as superscript numbers – for example, 10 to the power 2 is written as 10^2 – the small raised number 2 is the power.

A positive power means you multiply by that power of 10. Essentially this means that you need to multiply by 10 the same number of times as the power. For example, 1×10^3 has the power 3, so we multiply 1 by 10 three times:

$$1 \times 10 \times 10 \times 10 = 1000 = 1 \times 10^3$$

Table 1.4 Positive powers of 10

Number	Written	Often written
10	1×10^1	10
100	1×10^2	10^2
1000	1×10^3	10^3
10000	1×10^4	10^4
100000	1×10^5	10^5
1000000	1×10^6	10^6

When representing numbers that are smaller than 1 in standard form, you get negative powers (for example 1×10^{-1}). Essentially, this means that you need to divide by 10 the same number of times as the power. For example:

1×10^{-2} has the power –2, so we need to divide 1 by 10 twice:
$1 \div 10 \div 10 = 0.01 = 1 \times 10^{-2}$.

Table 1.5 Negative powers of 10

Fraction	Decimal	Written	Often written
$\frac{1}{10}$	0.1	1×10^{-1}	10^{-1}
$\frac{1}{100}$	0.01	1×10^{-2}	10^{-2}
$\frac{1}{1000}$	0.001	1×10^{-3}	10^{-3}
$\frac{1}{10000}$	0.0001	1×10^{-4}	10^{-4}
$\frac{1}{100000}$	0.00001	1×10^{-5}	10^{-5}
$\frac{1}{1000000}$	0.000001	1×10^{-6}	10^{-6}

Tip

When multiplying numbers in standard form, add the powers together and multiply the other numbers.

Tip

A number that you will use in Chemistry that is usually presented in standard form is Avogadro's number, which is 6.02×10^{23}.

Key term

Avogadro's number: The number of atoms, molecules or ions in one mole of a given substance.

A Worked examples

1 **A xylem vessel has a width of 0.072 mm. Write this width in standard form.**

Step 1 We know that $0.01 = 1 \times 10^{-2}$

Step 2 As the width is 0.072, replace the 1 with 7.2

Step 3 This gives an answer of $0.072\,\text{mm} = 7.2 \times 10^{-2}\,\text{mm}$

2 **Write the following wavelengths in standard form.**

a **Orange light (0.000 000 58 metres).**

Step 1 Write down 5.8

Step 2 The d.p. is now after the 5 (in 5.8); the d.p. has moved seven places to the right (from 0.000 000 58 to 5.8)

Step 3 The wavelength in standard form is $5.8 \times 10^{-7}\,\text{m}$

b **X-rays (0.000 000 000 195 metres).**

Step 1 Write down 1.95

Step 2 The d.p. is now after the 1 (in 1.95); the d.p. has moved 10 places to the right (from 0.000 000 000 195 to 1.95)

Step 3 The wavelength in standard form is 1.95×10^{-10} m

B Guided questions

1 **Calculate the number of atoms present in 2.3 g of sodium.**

Step 1 Calculate the amount in moles of sodium using

$$\text{Amount (in moles)} = \frac{\text{mass (g)}}{M_r} = \frac{2.3}{23} = 0.1$$

Step 2 One mole of atoms contains 6.2×10^{23} atoms. To find the number of atoms present in 2.3 g of sodium multiply the amount in moles by 6×10^{23}

2 **A species of bacterium divides every two hours. If there are 10 bacteria in the original population, how many bacteria would there be after 24 hours? Use the equation below and give your answer in standard form:**

Bacterial population = initial bacterial population × $2^{\text{number of divisions}}$

Step 1 Work out how many divisions will occur in 24 hours. Do this by dividing the mean division time by the total time.

$24 \div 2 = 12$

Therefore there are 12 divisions in 24 hours.

Step 2 Substitute the values into the equation.

Bacterial population = 10×2^{12} = ...

Key term

Mean: The mean is a type of average. Means are covered on pages 26-28.

C Practice questions

3 The surface of the Earth is divided into plates. In the North Atlantic Ocean, two of these plates meet. These plates are moving apart at about 25 mm per year.

How far apart will they move in five hundred thousand years? Give your answer in metres in standard form.

4 A species of bacterium divides every 5 hours. If there are 200 bacteria in the original population, how many bacteria would there be after 30 hours? Give your answer in standard form.

5 A copper atom has a diameter of 0.256 nm. A copper wire has a diameter of 0.044 cm.

- **a** Write the diameter of the atom and the wire in metres.
- **b** How many times wider is the copper wire than a copper atom? Give your answer to 2 decimal places in standard form.

6 An atom of hydrogen contains a proton and an electron. Calculate the mass of a hydrogen atom if a proton has mass 1.6725×10^{-24} g and an electron has mass 0.0009×10^{-24} g.

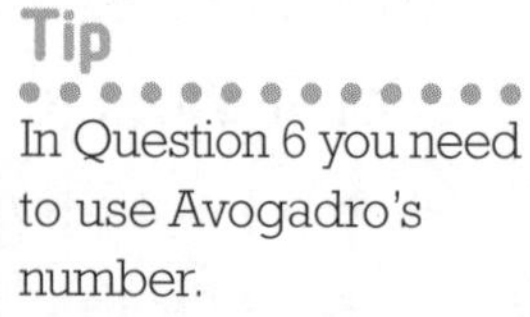

Tip

In Question 6 you need to use Avogadro's number.

Using a calculator with numbers in standard form

In your exam you may need to use standard form with a scientific calculator. Most calculators have a display of around nine numbers across the screen, which means very large numbers and very small numbers cannot be entered in normal form.

If, for example, we wanted to calculate $(2.99 \times 10^3) \times (4.1 \times 10^8)$ we would:

Step 1	key in 2.99
Step 2	press the '$\times 10^x$' key (on some calculators the '$\times 10^x$' key is labelled 'EXP')
Step 3	key in 3
Step 4	Press the multiply button, ×
Step 5	key in 4.1
Step 6	press the '$\times 10^x$' key
Step 7	key in 8
Step 8	press the '=' key to show the answer, 1.2259×10^{12}

To enter 2.99×10^{-3}, press the – or ± key before entering the number 3.

Tip

On many calculators, if you press the '=' key after entering a number in standard form, the number is displayed in normal form. Unfortunately, this does not work in reverse. However, pressing the 'ENG' button shows the number in engineering form, which is similar to, but not exactly the same as, standard form.

Ratios, fractions and percentages

Normal form (decimals), standard form, fractions and percentages are all numbers that we can key into a calculator. As all of these forms represent numbers, they can be changed from one form to another, as shown in Table 1.6.

Table 1.6 Numbers in different forms

Normal form	Standard form	Fraction	Percentage
0.03	3×10^{-2}	$\frac{3}{100}$	3%
0.5	5×10^{-1}	$\frac{1}{2}$	50%
3.7	3.7×10^{0}	$\frac{37}{10}$	370%
12.25	1.225×10^{1}	$12\frac{1}{4}$	1225%

In the following section we will look at fractions, ratios and percentages, which you will have to use in many calculations.

Key terms

Fraction: A number that represents part of a whole.

Numerator: The number on the top of the fraction.

Denominator: The number on the bottom of the fraction.

Common factor: A whole number that will divide into both the numerator and denominator of a fraction to give whole numbers.

Fractions

A fraction is part of a whole, and is expressed as a whole number divided by another whole number. The number on the top of the fraction is the numerator and the number on the bottom of the fraction is the denominator.

When using fractions, it is good practice to write each fraction in its simplest form, for example $\frac{5}{10}$ could also be written as $\frac{4}{8}$, $\frac{3}{6}$, or $\frac{2}{4}$; however the simplest form is $\frac{1}{2}$ so this should be used.

To find the simplest form of a fraction, divide the numerator and denominator (top number and bottom number) by the same whole number (a common factor), and carry on doing this until you are left with numerators and denominators that cannot be divided further to give whole numbers.

For example, in $\frac{2}{8}$ the numerator and the denominator can both be divided by 2 to give whole numbers, so $\frac{2}{8} = \frac{1}{4}$

Tip

If both the numerator and the denominator are even numbers, then the fraction is not in its simplest form.

In $\frac{9}{12}$ both the numerator and the denominator can be divided by 3 to give whole numbers, so $\frac{9}{12} = \frac{3}{4}$

3 and 4 cannot be further divided by the same number to give whole numbers, so $\frac{3}{4}$ is the simplest way of writing this fraction.

Calculating fractions (the traditional way)

Most scientific calculators have the ability to give answers as a fraction, such as $\frac{5}{24}$, although you can set your calculator to display a decimal number instead. To convert from a fraction to a decimal, you need to divide the numerator (top number) by the denominator (bottom number):

$$\frac{5}{24} = 0.208 \text{ (3 d.p.)}$$

Your calculator will have a button to do this calculation for you, often labelled S⇔D.

You need to be able to add, subtract, multiply and divide fractions.

Multiplying fractions

Multiplication is very straightforward. You multiply together the numbers on the top (numerators) and then the numbers on the bottom (denominators).

For example: $\frac{2}{3} \times \frac{3}{4} = \frac{6}{12}$

We can simplify the fraction by dividing the numerator and denominator by 6 to give $\frac{1}{2}$.

Dividing fractions

To divide fractions, we invert the divisor (the second fraction) and multiply.

For example: $\frac{3}{4} \div \frac{7}{8} = \frac{3}{4} \times \frac{8}{7} = \frac{24}{28}$

We can simplify the fraction by dividing the numerator and denominator by 4 to give $\frac{6}{7}$.

Adding and subtracting fractions

It is easy to add or subtract fractions if they have the same denominator.

Example 1: $\frac{2}{7} + \frac{4}{7} = ?$

Here, the common denominator is 7, so $\frac{2}{7} + \frac{4}{7} = \frac{6}{7}$

Example 2: $\frac{7}{12} - \frac{5}{12} = ?$

Here, the common denominator is 12, so $\frac{7}{12} - \frac{5}{12} = \frac{2}{12}$

We can simplify the fraction by dividing the numerator and denominator by 2 to give $\frac{1}{6}$.

If we are not given a common denominator we need to find one. Any two (or more) fractions will share a common denominator that will allow us to add or subtract them. To find a common denominator you can multiply the two denominators together.

For example:

$$\frac{1}{3} + \frac{1}{4} = ?$$

Tip

If you come across a fraction where the numerator is bigger than the denominator, such as $\frac{9}{6}$, this means that it can be simplified as a whole number and a fraction. In this example you have $\frac{6}{6}$ plus another $\frac{3}{6}$, or $1\frac{3}{6} \equiv 1\frac{1}{2}$.

In this example, we can't add them together as is. However, if we multiply the denominators together, we get $3 \times 4 = 12$. To make sure the fractions stay equivalent (the same) when we change the denominator, we also have to multiply the numerator by the same number. This is because you can see that $\frac{1}{3}$ is not the same value as $\frac{1}{12}$. We multiplied the denominator (3) by 4 to get 12, so we need to multiply the numerator (1) by 4 as well. If we do this to both fractions we get:

$$\frac{4}{12} + \frac{3}{12} = ?$$

Now we can add them together to get $\frac{7}{12}$.

Calculating fractions (the easy way)

In every GCSE Science exam you are expected to be able to use a scientific calculator. Learning how to use a calculator to work out fractions is straightforward.

The instructions below tell you how to enter the fraction $\frac{3}{4}$ and then add $\frac{1}{2}$.

- look for the fraction button; the symbol will appear on the screen when you press it
- press the number 3 button; the 3 is displayed as the numerator
- press 'down' on the navigation button when you are ready to enter the denominator
- press the number 4 button
- press 'right' on the navigation button so the fraction, $\frac{3}{4}$, is on the screen.
- press +
- enter the fraction $\frac{1}{2}$ in the same way as you entered $\frac{3}{4}$, finishing by pressing 'right' on the navigation button
- press =
- the display should show the answer: $\frac{5}{4}$ or $1\frac{1}{4}$.
- press S⇔D to see the answer displayed as a decimal (1.25).

Mixed fractions are fractions that have a whole number bit and a fraction bit. For example, $2\frac{1}{2}$ is a mixed fraction.

The instructions below tell you how to enter the mixed fraction $2\frac{1}{2}$ and then multiply by $3\frac{3}{4}$.

- find and press the mixed fraction symbol – this is usually above the fraction button; you need to press the shift button, then the fraction button
- you can now enter a mixed fraction
- press the number 2 button, and then press 'right' on the navigation button
- press the number 1 button, and then press 'down' on the navigation button
- press the number 2 button, and then press 'right' on the navigation button
- the display now shows $2\frac{1}{2}$
- press ×

Tip

If the common denominator you get by multiplying both denominators together is very high, see if you can simplify it by reducing it to a smaller number. For example, the fraction sum $\frac{24}{56} + \frac{8}{56}$ might be more easily understood as $\frac{3}{7} + \frac{1}{7}$, as both fractions share the same multiple of 8. If you divide both fractions by 8, you get a much simpler sum. You can also do this simplification after completing the sum, if you prefer.

- enter the mixed fraction $3\frac{3}{4}$ in the same way as you entered $2\frac{1}{2}$
- press =
- the display should show $9\frac{3}{8}$ or $\frac{75}{8}$
- press S⇔D to see the answer displayed as a decimal (9.375).

Once you know how to enter fractions in your calculator, you can add, subtract, multiply and divide them easily.

Tip

Remember that using a calculator is a skill and skills are not learned overnight. They need practice!

C Practice questions

1 Complete these calculations using a calculator.

a $1\frac{1}{4}+3\frac{5}{8}$

b $2\frac{2}{3}+4\frac{5}{6}$

c $7\frac{5}{12}-6\frac{1}{4}$

d $3\frac{2}{5}-4\frac{7}{10}$

2 Alloys are mixtures of metals. Gold is often alloyed with copper to make it harder and more hard-wearing.

9-carat gold contains $\frac{3}{8}$ pure gold and $\frac{5}{8}$ copper by mass.

Calculate the mass of a piece of 9-carat gold if it contains 95 g of copper.

Tip

You know that $\frac{5}{8}$ of the gold is copper, so use this to first find the mass of $\frac{1}{8}$, and then use this amount to find the mass of $\frac{8}{8}$ (in other words, all) of the 9-carat gold.

Percentages

Like fractions, percentages represent part of a whole. Unlike fractions, they are expressed in the form of a number followed by the percentage symbol %, which means 'divided by 100' or 'out of 100'.

For example: $\frac{1}{4}=\frac{25}{100}=25\%$

To convert a fraction into a percentage, divide the numerator (top number) by the denominator (bottom number) and multiply by 100.

It can be difficult to compare fractions when they have different denominators. For example, it is not easy to say whether $\frac{3}{10}$ is bigger or smaller than $\frac{4}{11}$ without doing some calculations. Percentages solve that problem as a percentage is a fraction of 100.

Table 1.7 Some common percentages, decimals and fractions

Fraction	$\frac{1}{20}$	$\frac{1}{10}$	$\frac{1}{4}$	$\frac{1}{2}$	$\frac{3}{4}$	1
Decimal	0.05	0.10	0.25	0.50	0.75	1.00
Percentage	$\frac{1}{20}=\frac{5}{100}$ $=5\%$	$\frac{1}{10}=\frac{10}{100}$ $=10\%$	$\frac{1}{4}=\frac{25}{100}$ $=25\%$	$\frac{1}{2}=\frac{50}{100}$ $=50\%$	$\frac{3}{4}=\frac{75}{100}$ $=75\%$	$\frac{1}{1}=\frac{100}{100}$ $=100\%$

To calculate a *percentage change* – this may be an increase or a decrease – use the following equation.

$$\text{Percentage change} = \frac{\text{change in value}}{\text{original value}} \times 100$$

Other types of calculation that involve percentages include percentage yield and atom economy calculations. To calculate these, you need to recall and use the equations shown below.

$$\text{Percentage yield} = \frac{\textit{actual}\text{ yield}}{\text{theoretical yield}} \times 100$$

$$\%\text{ atom economy} = \frac{\text{molecular mass of desired product}}{\text{sum of molecular masses of all reactants}}$$

A Worked examples

1 **An investigation was carried out into the effect of changing NaCl concentration on the mass of a sample of carrot in solution. The sample of carrot lost 3 g of its total mass of 10 g. What percentage of its mass did the carrot lose?**

Step 1 In this case, the 'whole' is 10 g and the 'part' is 3 g, so the carrot lost $\frac{3}{10}$ of its mass.

Step 2 To convert this fraction to a percentage, divide the numerator (3) by the denominator (10) and multiply by 100:

$$\frac{3}{10} \times 100 = 30\%$$

2 **A 2.4 g sample of iron ore contains 1.8 g of Fe_2O_3. What percentage of the iron ore is Fe_2O_3?**

Step 1 Express the quantity as a fraction: $\frac{1.8}{2.4}$

Step 2 Multiply by 100: $\frac{1.8}{2.4} \times 100 = 75\%$

B Guided questions

1 **Calculate the percentage of nitrogen in $Ca(NO_3)_2$.**

Step 1 In this example you first need to find the relative formula mass (M_r) of $Ca(NO_3)_2$

$$M_r = 40 + (14 \times 2) + (16 \times 6) = \text{.................}$$

Step 2 There are two nitrogen atoms and so the mass of nitrogen in $Ca(NO_3)_2$ is $14 \times 2 =$ ....

Step 3 Express the quantity as a fraction

$$\frac{\text{mass of nitrogen}}{M_r} = \text{.................}$$

Step 4 Multiply by 100 to express as a percentage.

2 **In a day, 4000 kJ of light energy from the Sun falls on a plant. The plant converts 52 kJ of this energy into photosynthetic products.**

Calculate how efficient this energy transfer is, giving your answer as a percentage.

To calculate the percentage efficiency of this energy transfer, divide the amount of energy in the photosynthetic products by the total energy falling on the plant, and then multiply the answer by 100.

Step 1 Efficiency of energy transfer = ... ÷ .... × 100

Step 2 Efficiency of energy transfer = ...

Tip

Remember the rules for converting fractions, decimals and percentages.

C Practice questions

3 A car is supplied with 30 MJ of chemical energy. Of this, 21 MJ are wasted and the rest is converted into useful kinetic energy.

What percentage of the input energy is:

a wasted

b converted into useful energy?

4 An investigation was carried out into the transfer of biomass through a moorland ecosystem. The results were used to draw the following food chain.

Heather 300 000 kJ → Grouse 19 000 kJ → Fox 2100 kJ

Calculate the efficiency of the transfers below. In each case, represent your answer as both a percentage and a fraction in its simplest form.

a the heather and the grouse

b the grouse and the fox

5 Calculate the percentage by mass of

a Hydrogen in $Ca(OH)_2$

b Potassium in $K_2Cr_2O_7$

c Nitrogen in $(NH_4)_2SO_4$

Ratios

A ratio expresses a relationship between quantities. It shows how many of one thing you have relative to how many of one or more other things. In ratios, the numbers are separated by a colon (:).

For example, suppose that when two plants of a certain species are crossed, eight offspring with red petals are produced for every four offspring with purple petals. The ratio of plants with red petals to those with purple petals is therefore 8 : 4.

In another example, a step-up transformer may have a turns ratio $N_s:N_p$ of 3:1, which means that there are three times as many turns on the secondary coil N_s as there are on the primary coil N_p.

Sometimes it is easier simply to express a ratio as a whole number, as opposed to a fraction. In the example above we might say 'the turns ratio in the step-up transformer is 3'.

For a ratio to be valid, the quantities being compared must be *of the same unit*. So, a ratio, even when expressed as a whole number, a fraction or decimal, does not usually have a unit. Ratios are similar to fractions; they can (and should) both be simplified by finding common factors.

Tip

A factor is a number that divides exactly into another number.

Ratios are also used to show direct proportion. Look at the first two rows of Table 1.8. When we double (or triple, or quadruple) the mass, we do the same thing to the volume – that's what direct proportion means.

Table 1.8 $m:V$ ratios

Mass, m (g)	10	20	30	40
Volume, V (cm^3)	2	4	6	8
Ratio $m:V$	10:2 = 5:1	20:4 = 5:1	30:6 = 5:1	40:8 = 5:1

The last row in the table also shows that the ratio $m:V$ is always the same (in this case 5:1). The constant ratio is a test for direct proportion. But ratios are like fractions. In this case the ratio $m:V$ (or, if you prefer, the fraction $\frac{m}{V}$) is 5.

From the definition of density ρ you know that $\rho = \frac{m}{V}$. So, the ratio needs a unit; the unit for density is g/cm^3.

Ratios are used in Chemistry in many calculations, for example: in working out empirical formulae; in calculating reacting masses; and in balancing equations.

A Worked examples

1 In a genetic cross, the predicted ratio of offspring is 3 long-haired : 1 short-haired. If there were 20 offspring, how many offspring would you expect to have long hair and how many would have short hair?

Step 1 Add the numbers in the ratio together: $3 + 1 = 4$

Step 2 Divide the total number of offspring by the number found in Step 1.

$20 \div 4 = 5$

This is how many each '1' in the ratio represents.

Step 3 Multiply each number in the ratio by the value found in Step 2.

Therefore, we expect there to be:

$3 \times 5 = 15$ long-haired offspring

$1 \times 5 = 5$ short-haired offspring

2 The total input energy to a simple motor is 3000 J. The useful output energy is 1.8 kJ.

Calculate the motor's efficiency.

Step 1 Write down what we mean by efficiency:

$$\text{efficiency} = \frac{\text{useful output energy}}{\text{total input energy}}$$

Step 2 Substitute the numbers: efficiency $= \frac{1800\,\text{J}}{3000\,\text{J}}$

Notice that in this question we changed the 1.8 kJ to 1800 J. This is because when we calculate a ratio, both numbers must have the same unit.

Step 3 Do the calculation: efficiency = 0.6

B Guided questions

1 A compound contains 0.050 moles of phosphorus and 0.125 moles of oxygen atoms. What is its molecular formula?

Step 1 Write down the elements present and the moles of each underneath.

P : O

.... : 0.125

Step 2 To find the simplest ratio divide by the smaller of the number of moles.

$\frac{0.050}{0.050} : \frac{0.125}{\ldots\ldots\ldots}$

Sometimes you may not get a whole number ratio at this stage, and often multiplying by 2 or another number is necessary.

2 A genetic cross was carried out to determine the expected offspring from breeding two fish together. In this species of fish, red stripes are dominant to orange stripes. One of the fish was heterozygous and had red stripes, and the other was homozygous and had orange stripes.

Use a Punnett square diagram to determine the expected ratio of offspring that have orange stripes to those that have red stripes.

Step 1 Use R for the dominant allele, and r for the recessive allele.

The red-striped parent has a genotype of Rr.

The orange-striped parent has a genotype of rr.

Step 2 This gives the following genetic cross:

Parents: Rr rr

Gametes: R r r r

	r	r
R		
r		

Step 3 Expected ratio of offspring = ...

C Practice questions

3 The current flowing in a resistor is measured when the voltage across it is changed and the results are recorded in a table, like the one shown below.

By calculating a suitable ratio, show that the voltage is directly proportional to the current.

Voltage, V (V)	3.2	4.0	4.8	5.6	6.4	7.2
Current, I (A)	0.20	0.25	0.30	0.35	0.40	0.45
Ratio						

4 Write the empirical formula of the following compounds

a $C_{16}H_{20}N_8O_4$

b $Na_2S_2O_3$

c $C_6H_{12}O_6$

d P_4O_{10}

Balancing equations

In a balanced chemical equation, the substances are all in ratio to each other and this is shown by the numbers in front of each formula in the balanced symbol equation. For example, 2 moles of magnesium react with 1 mole of oxygen to produce 2 moles of magnesium oxide.

$$2Mg + O_2 \rightarrow 2MgO$$

The ratio is

$$2\text{ moles Mg} : 1\text{ mole } O_2 : 2\text{ moles MgO}$$

Or in the reaction

$$2Al + 6HCl \rightarrow 2AlCl_3 + 3H_2$$

The ratio between aluminium and hydrogen is

$$2\text{ moles Al} : 3\text{ moles } H_2$$

The ratio between aluminium and hydrochloric acid is

$$2\text{ moles Al} : 6\text{ moles HCl}$$

which simplifies to

$$1\text{ mole Al} : 3\text{ moles HCl}$$

Ratios can be used in calculating the number of moles that react together in a reaction.

A Worked example

1 **In the reaction $4Al + 3O_2 \rightarrow 2Al_2O_3$**

a **How many moles of aluminium are needed to produce 0.76 moles of aluminium oxide?**

Step 1 Write down the ratio between the two substances using the equation and simplify.

$Al : Al_2O_3$

4 : 2

2 : 1

Step 2 Apply this ratio to the 0.76 moles of Al_2O_3

There are twice as many moles of aluminium so multiply by 2

$0.76 \times 2 = 1.52$ mol

b **How many moles of aluminium oxide are produced from 0.2 moles of aluminium?**

Step 1 Write down the ratio between the two substances using the equation and simplify.

$Al : Al_2O_3$

4 : 2

2 : 1

Step 2 Apply this ratio to the 0.2 moles of aluminium.

There is half as many moles of aluminium oxide so divide by 2

$\frac{0.2}{2} = 0.1$ mol

B Guided question

1 **In the reaction $N_2 + 3H_2 \rightarrow 2NH_3$ how many moles of nitrogen are needed to react fully with 0.4 moles of hydrogen?**

Step 1 Write down the ratio between the two substances using the equation and simplify.

$N_2 : H_2$

1 : 3

Step 2 Apply this ratio to the 0.4 moles of hydrogen.

There is three times as much hydrogen as nitrogen so divide the moles of hydrogen (0.4) by 3.

....

C Practice question

2 In the reaction $2Cu(NO_3)_2\ (s) \rightarrow 2CuO\ (s) + 4NO_2\ (g) + O_2\ (g)$

a How many moles of O_2 are produced from 4 moles of $Cu(NO_3)_2$?

b How many moles of NO_2 are produced from 0.6 moles of $Cu(NO_3)_2$?

Estimating results

★ **Not explicitly required for WJEC/Eduqas Science Double Award.**

When carrying out calculations, it can be useful to estimate the answer first. Estimates can mean that obvious mistakes are spotted quickly. Estimates can be simple guesses based on your experiences, or they can be based on quick calculations. For example, if somebody told you that the speed of an athlete is 100 m/s, a quick estimate shows you that they must be wrong as the world record for 100 m is just under 10 seconds.

Estimates can also help you see if you have entered the wrong number on your calculator, or divided instead of multiplying, as your estimate will show that your answer is clearly wrong. You can then re-check the calculation and correct your mistake.

When estimating, it is meant to be quick. This means you need to make the calculations as easy as possible. The best way to do this is to round each given value to the nearest ten, hundred or other convenient whole number. While your answer will not be the 'correct' number, it will be an approximate estimation of it.

The first step, when estimating, is to convert the numbers to 1 significant figure (s.f.). So, for example, instead of multiplying by 112, we would multiply by 100. Instead of dividing by π (which is roughly 3.14), we would divide by 3. Rather than divide by 19.3, we would divide by 20, and so on.

Tip

While estimating is a useful skill that can help you check if a calculation is correct, in an exam it is important to use your calculator to find the value precisely and write this as the answer.

Tip

If you are unsure of how to round to 1 significant figure see pages 23–25.

A Worked example

An area of the Amazon rainforest has suffered intense deforestation. The area affected was 33 km long by 1.89 km wide. Estimate the total area affected.

Step 1 For a quick estimation of the area, round both given values to make the calculation more straightforward.

33 km rounds down to 30 km.

1.89 km rounds up to 2 km.

Step 2 Perform the calculation with the rounded values.

This gives an estimated area of $30\ km \times 2\ km = 60\ km^2$

When a calculator is used to find the area using the values stated in the question, it gives an answer of: $33\ km \times 1.89\ km = 62.37\ km^2$

Clearly, the estimated and actual values are different: $60\ km^2$ is not the correct answer, but our estimate is close.

Tip

Estimates can also help you spot obvious mistakes in your answers. For example, in this question, if you had pressed ÷ instead of × on the calculator, we would have got $17.46\ km^2$, which looks wrong just from looking at it. We could then go back and correct the mistake.

B Guided question

1 **In a chromatography experiment, a student found that a compound moved a distance of 8.2 cm and the solvent moved a distance of 19.6 cm. Estimate the R_f value and use it to decide if the compound is P, which has R_f value 0.4, or Q, which has R_f value 0.2.**

Step 1 Write down the equation used to calculate R_f

$$R_f = \frac{\text{distance moved by compound}}{\text{distance moved by solvent}}$$

Step 2 Round each distance to one significant figure

Compound distance = 8 solvent distance = ...

Step 3 Estimate the R_f value.

....

C Practice questions

2 Light travels at 3.0×10^8 m/s.

Estimate how long light would take to travel 400 000 km to the Moon and 400 000 km back again. Give your answer to the nearest second.

3 During the day, a person's blood glucose concentration decreased from 6.3 mmol/L to 3.9 mmol/L. To estimate the percentage change in blood glucose concentration, a researcher did the following calculation:

$\frac{3}{6} \times 100\% = 50\%$

Is this the best estimate the researcher could have made? Explain your answer.

Using sin and $\sin^{-1}$ keys

Some students taking GCSE Science must be able to use the sin and $\sin^{-1}$ keys on the calculator to solve problems on refraction. The example below shows how.

★ **Only required for Edexcel International GCSE Science Double Award.**

A Worked example

The diagram shows a ray of light passing through a rectangular glass prism.

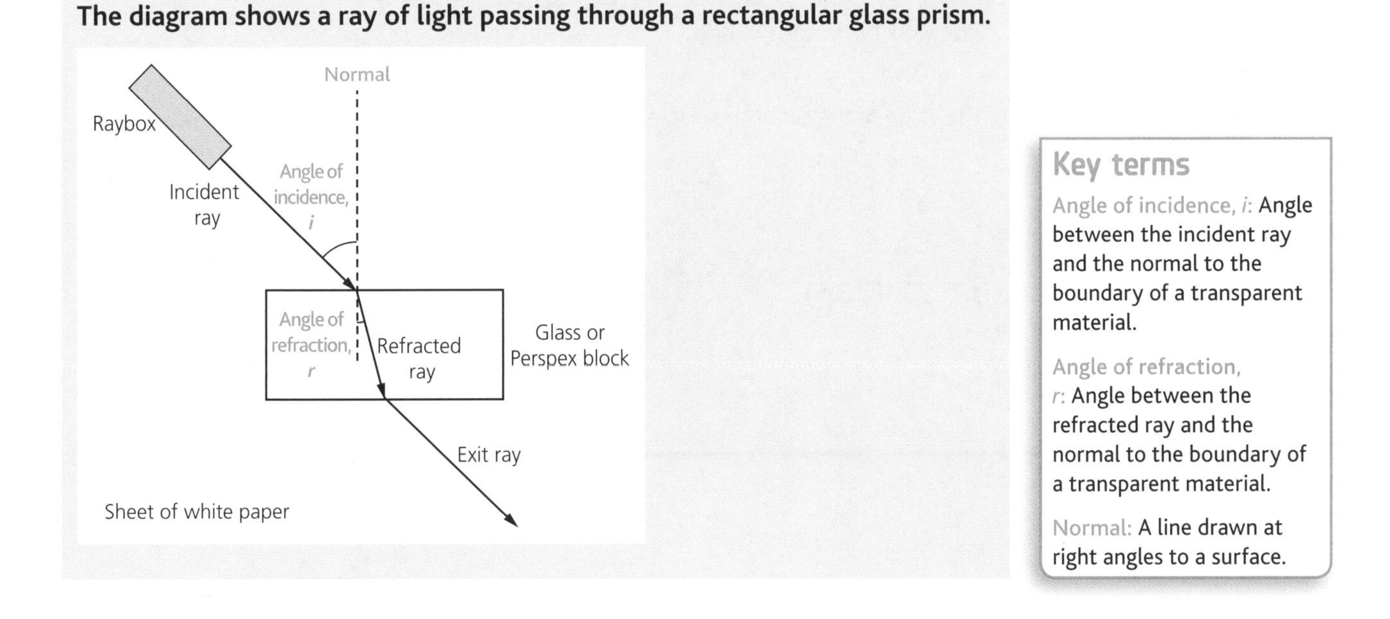

Key terms

Angle of incidence, *i*: Angle between the incident ray and the normal to the boundary of a transparent material.

Angle of refraction, *r*: Angle between the refracted ray and the normal to the boundary of a transparent material.

Normal: A line drawn at right angles to a surface.

a **If the angle of incidence *i* in air is 40° and the angle of refraction *r* in glass is 25°, calculate the refractive index of the glass, *n*, giving your answer to 2 d.p.**

Step 1 The mathematical relationship between i and r is $n = \frac{\sin i}{\sin r}$

Step 2 Substituting the numbers for i and r gives $n = \frac{\sin 40}{\sin 25}$

Step 3 To calculate n, using a calculator, do the following:

Action	Display shows
press the sin button	sin(
enter 40 and close the bracket	sin(40)
press the ÷ button	sin(40) ÷
enter sin(25) just as you entered sin(40)	sin(40) ÷ sin(25)
press =	1.52096

So, the refractive index is 1.52 (2 d.p.).

b **Using your answer to part a, find the value of *i* when the angle of refraction is 40°, giving your answer to 2 d.p.**

Step 1 To find an angle such as the value of i, we need to use the $\sin^{-1}$ key.

Step 2 Again, we will use the equation $n = \frac{\sin i}{\sin r}$

Step 3 If we multiply both sides of the equation by $\sin r$, we get $\sin i = n \times \sin r$.

Step 4 The first step to finding i is to calculate $\sin i$.

Step 5 So, we have to calculate $n \times \sin r$:

Action	Display shows
enter 1.52 × sin	1.52×sin(
enter 40)	1.52×sin(40)
press =	0.977037
So, sin*i* = 0.977037 and we can work backwards to find *i*:	
Press SHIFT and then sin	sin⁻¹(
The calculator is looking for the value of the sin of the angle, so that it can calculate the angle.	
* press ANS	sin⁻¹(ANS
close the bracket and press =	77.6977

So, the angle of incidence is 77.70° (2 d.p.).

Key term

Refractive index: The ratio $\sin i : \sin r$.

Tip

If your calculator does not have an ANS button, you could re-enter the number 0.977037 at the stage marked with the asterisk. Then close the bracket and press =.

B Guided question

1 The diagram shows refraction of light when the angle of incidence in glass is equal to the critical angle, *c*.

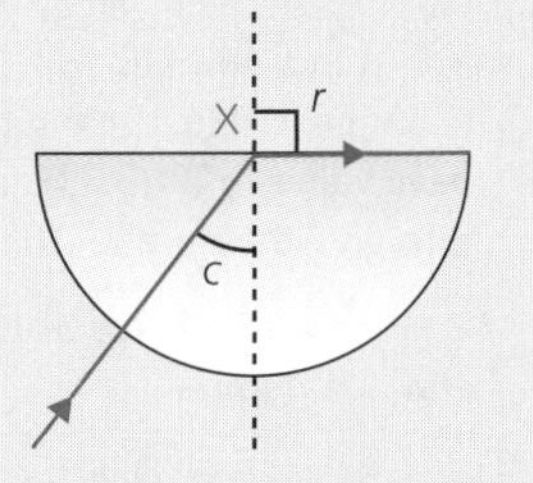

If the refractive index of the glass is 1.52, calculate the value of *c*, giving your answer to 1 d.p.

Step 1 $n = \frac{1}{\sin c} \Rightarrow c = \sin^{-1}\left(\frac{\dots\dots}{\dots\dots}\right)$

Step 2 So, $c = \dots\dots\dots\dots^\circ = \dots\dots\dots\dots^\circ$ (to 1 d.p.).

> **Key term**
>
> Critical angle: The angle of incidence in an optically dense medium when the angle of refraction in air is 90°.

C Practice questions

2 The angle of incidence in glass that gives an angle of refraction in air of 90° is called the critical angle.

Show that the refractive index of glass is 1.56 (2 d.p.) if the critical angle is 40°.

3 Using the information given in the diagram, calculate the refractive index of medium B with respect to medium A.

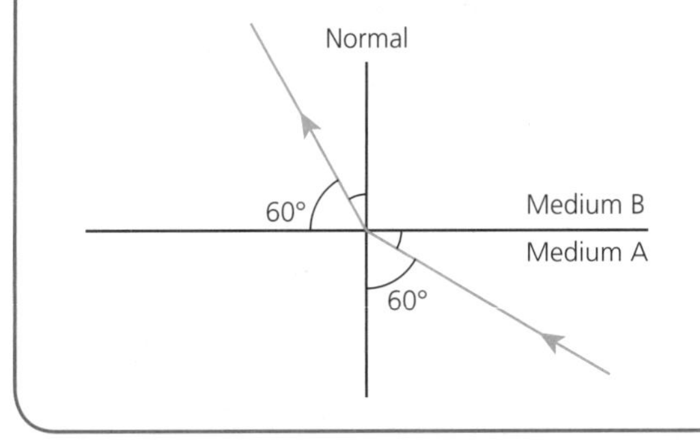

> **Tip**
>
> For advice on rearranging the subject of equations, see page 52.

Handling data

Significant figures

Significant figures can be a complex topic, but there are some general rules for using them. There are exceptions to the rules outlined below, but these are unlikely to feature in a GCSE exam. In general, *all* digits are significant figures except in the instances below:

- Leading zeros, for example zeros before a non-zero digit. For example, 0.07 has two leading zeros, and these are not significant figures. 0.07 only has one significant figure (7). The zeros are written to make the place value correct.

> **Tip**
>
> The term 'significant figures' can often be shortened to 's.f.' or 'sig fig', but they mean the same thing.

- Zeros after a non-zero digit if they are due to rounding or used to indicate place value. For example, a value that is rounded to the nearest hundred (for example 600 g) has two trailing zeros, which are not significant, and so it has only one significant figure (6). A value that is exactly 600 g, however, would have three significant figures, and the zeros in this case would be significant.
- Spurious digits, for example digits that make a calculated value appear more precise than the original data used in the calculation. For example, suppose that one side of a square was measured with a ruler to be 13.1 cm long. Using this measurement to calculate the area of the square gives a value of 171.61 cm^2 (13.1 × 13.1). The ruler measured only to three significant figures, while the answer appears to contain five significant figures. This means that the last two digits (6 and 1) are spurious and should not be included in the final answer. Therefore, the result should be rounded to 172 cm^2 (three significant figures as in the original measurement).

When two or more different pieces of measuring apparatus are used, calculated results should be reported to the limits of the least accurate measurement. This means that you should use the number of significant figures that the *least* accurate piece of apparatus measures. This will usually mean using the same number of decimal places as the least accurate piece of apparatus.

Giving the correct number of significant figures in calculations is important because it signifies the degree of precision.

Suppose a balance reads to the nearest 100 g:

- If the true mass on the balance was larger than 2450 g but less than 2500 g, the balance would round up to the nearest 100 g and give a reading of 2500 g.
- If the true mass was bigger than 2500 g but less than 2550 g, the balance would round down to the nearest 100 g and also give a reading of 2500 g.
- For all masses between 1000 g and 9900 g, this balance would give a figure to 2 s.f.
- For this balance, the last two digits will always be zero.

Suppose we look at a new balance capable of reading to the nearest 10 g and measure the same mass, and we get a reading of 2470 g.

- We would know for sure that the mass (M) can be estimated as $2465\text{ g} \leqslant M < 2475\text{ g}$.
- For all masses between 1000 g and 9990 g, this balance would give a figure to 3 s.f.
- For this balance, only the final digit would always be zero.

Suppose this second balance gave a reading of 2500 g, like the first balance. Could we say that the two balances gave the same information? The answer is no.

- The first balance can only tell us that $2450\text{ g} \leqslant M < 2550\text{ g}$, although the reading is 2500 g (2 s.f.).
- The second balance is tell us that $2495\text{ g} \leqslant M < 2505\text{ g}$, although the reading is also 2500 g (3 s.f.).

It is clear that the number of significant figures tells us something of the degree of precision in the instrument being used. See pages 112–113 for more detail on precision, accuracy and resolution.

Key terms

Leading zero: A zero before a non-zero digit, for example 0.6 has one leading zero.

Place value: The value of a digit in a number, for example in 926, the digits have values of 900, 20 and 6 to give the number 926.

Trailing zeros: Zeros at the end of a number.

Spurious digits: Digits that make a calculated value appear more precise than the data used in the original calculation.

Tip

Some answers may be recurring. For example, the calculator display shows 9.652652652… If this happens, use the recurring form of the number when writing the number to so many significant figures. So, 9.652652652… is 10 (to 1 s.f.), 9.7 (to 2 s.f.), 9.65 (to 3 s.f.), 9.653 (to 4 s.f.) and so on. You might find it useful to look again at the section on decimal places on page 6 – it is very similar to significant figures.

Tip

When answering mathematical questions, look to the data (numbers) in the question that are given to the least number of significant figures. Your final answer should have the same number of significant figures, unless the question tells you otherwise.

A Worked examples

1 **Identify how many significant figures the number 0.0304 has.**

Step 1 Identify the first non-zero digit from the left. This is 3.

Step 2 The two zeros to the left of 3 are leading zeros, and therefore not significant.

Step 3 The other three digits (3, 0 and 4) are significant.

Step 4 So this number has three significant figures.

2 **The mass of an adult human brain was recorded as 1368 g. Write this mass to two significant figures.**

Step 1 The first significant digit is 1, and the digit immediately to its right, 3, is the second significant digit. These are the two significant digits we need to use to work out our answer.

Step 2 To work out the answer, we need to determine if we can use 3 as the second significant figure (for example an answer of 1300) or will need to round it up to 4 (an answer of 1400). In order to work out if we should round up or down, look at the digit immediately to the right of 3.

Step 3 In this case the digit is 6, therefore we need to round 1368 up to 1400.

Step 4 Therefore, the mass is 1400 g to two significant figures. This could also be written as 1.4×10^3 g in standard form (see pages 8-11 for more detail on standard form).

B Guided question

1 **A current of 1.4 A flows through a resistor of 6.8 Ω.**

Calculate the voltage across the resistor, giving your answer to an appropriate number of significant figures.

Step 1 Write equation for Ohm's Law: $V = I \times R$

Step 2 Substitute for I and R: $V =$ ....

Step 3 Do the arithmetic: $V =$ .... volts

Step 4 Number of s.f. in data in question is two.

Step 5 Give answer to appropriate number of s.f.: $V =$ .... volts.

C Practice questions

2 During an investigation into the rate of an enzyme catalysed reaction, a balance was used to measure 71.6 g of product, which was produced in 10.5 hours. This mass was used to calculate a rate of reaction of 6.819 g/hour. Write this answer to the correct number of significant figures.

3 How much heat energy is needed to raise the temperature of 2.55 kg of water by 12.2 °C if the specific heat capacity of water is 4200 J/kg°C? Give your answer to an appropriate number of significant figures.

4 In an experiment the theoretical yield to prepare some copper(II) sulfate crystals was 2.85 g.

$$CuO + H_2SO_4 \rightarrow CuSO_4 + H_2O$$

Only 2.53 g of copper(II) sulfate was obtained. Calculate the percentage yield of copper(II) sulfate in this experiment and give your answer to two significant figures.

Finding arithmetic means

The arithmetic mean (often denoted by $\bar{x}$ in tables or equations) is an 'average', and is calculated by adding all the individual values in a data set together and dividing by the total number of values used.

Key term

Arithmetic mean: The sum of a set of values divided by the number of values in the set – it is sometimes called the average or mean.

The mean is the most commonly used average at GCSE, and the one you will normally use in practical investigations. For example, you might use it when investigating the effect of light intensity on the number of bubbles produced by pond weed. Each light intensity can be repeated three times and a mean number of bubbles found. By taking the mean, we hope that the numbers that are too big cancel out the numbers that are too small.

Using the mean when averaging data does have some disadvantages because it can be skewed by extreme (or outlier) results. An example of this is shown in Table 1.9.

Key term

Outlier: A data point that is much larger or smaller than the nearest other data point.

Table 1.9 Disadvantages of averaging data

Temperature (°C)	20	20	20
Rate of reaction of protease (1 / time taken)	0.1	0.2	0.9

In this data set, 0.9 is very different from the other values so it seems highly likely that this value is an outlier. This means that it can be excluded when calculating the mean:

$$\frac{(0.1 + 0.2)}{2} = 0.15$$ (2 being the number of data values used to calculate the mean)

If 0.9 had been included, this would give a mean value of:

$$\frac{(0.1 + 0.2 + 0.9)}{3} = 0.4$$

This second value is much larger than the mean calculated when not using the outlier, and so is not as representative of the data.

Tip

When outliers are present in a data set, it may be more appropriate to use another type of average (such as the median or mode; see page 28) or you can just discard the outlier values before calculating the mean.

The arithmetic mean is found by adding together all the values and dividing by the total number of values. It may be referred to as the 'average' or simply as the mean.

A Worked example

The results from the titration between hydrochloric acid and 25.0 cm³ of sodium hydroxide were recorded in the table below. Concordant results are those that are within ±0.10 cm³ of each other.

Use the concordant results to calculate the mean volume of hydrochloric acid required to neutralise the 25.0 cm³ of sodium hydroxide.

	Titration 1	Titration 2	Titration 3	Titration 4
Final burette reading/cm³	26.10	25.20	25.45	25.15
Initial burette reading/cm³	0.00	0.10	0.00	0.00
Volume/cm³	26.10	25.10	25.45	25.15

Step 1 Determine the concordant results.

The results of titration 1 and 3 are not concordant and are not used to calculate the mean titre.

Step 2 Mean titre $= \frac{25.10 + 25.15}{2} = 25.13\,\text{cm}^3$

B Guided question

1 **Five students independently measure the resistance of a length of wire. They obtain these results:**

2.1 Ω, 2.2 Ω, 0.5 Ω, 1.9 Ω, 1.8 Ω

a **Identify the outlier.**

Step 1 The outlier is: ....

b **Calculate the mean of the other four resistances.**

Step 1 The sum of the other results is 2.1 + ... + ... + ... = ...

Step 2 The mean resistance is .... ÷ 4 = ... Ω

C Practice questions

2 The table below shows the results of an investigation into light intensity at three different sites. Complete the table to show the mean light intensity at site B.

Site	Light intensity (lumen)			
	1	2	3	Mean
A	1900	1800	1950	1883
B	1500	1600	1700	
C	1200	1350	1250	1267

3 The table shows the level of nitrates present in water from two different rivers taken at four different points along the river. Nitrate levels below 10 mg/l are considered safe to drink.

River	Nitrate level (mg/l)				
	Point 1	Point 2	Point 3	Point 4	Point 5
A	14	13	11	9	8
B	8	9	10	11	9

Calculate the mean nitrate level in each river. Which river is safest to drink from?

4 The generally accepted value for the specific heat capacity of water is 4.2 J/g°C.

A group of 10 students measure the specific heat capacity of water and their mean result is the generally accepted value.

Nine of the students' results, in J/g°C, are:

4.1, 4.2, 4.2, 4.3, 4.3, 4.1, 4.2, 4.0, 4.1

Calculate the value for the specific heat capacity obtained by the tenth student.

Calculating weighted means

A weighted mean is where some values contribute more to the mean than others.

Relative atomic mass (A_r) is a weighted mean of isotopic masses. The relative atomic mass of an element can be calculated from the relative isotopic masses of the isotopes (which are the same as the mass numbers) and the relative proportions in which they occur (abundance).

$$\text{Relative atomic mass} = \frac{\text{sum of (mass} \times \text{abundance) for all isotopes}}{\text{total abundance}}$$

A Worked example

Calculate the relative atomic mass of rubidium to two decimal places.

	Relative isotopic mass	Abundance /%
^{85}Rb	85	72.15
^{87}Rb	87	27.85

Step 1 Find the total abundance.

$72.15 + 27.85 = 100$

Step 2 Calculate the relative atomic mass.

$$\text{Relative atomic mass} = \frac{(85 \times 72.15) + (87 \times 27.85)}{100} = 85.56$$

B Guided question

1 **The element magnesium contains 79% ^{24}Mg, 10% ^{25}Mg and 11% ^{26}Mg. Calculate the relative atomic mass of magnesium to one decimal place.**

Step 1 First find the total abundance

$79 + 10 + 11 =$ ....

Step 2 Then calculate relative atomic mass by multiplying each relative isotopic mass value by its abundance and dividing by the total abundance.

$$\text{Relative atomic mass} = \frac{(79 \times 24) + (10 \times \text{.........}) + (\text{.........} \times \text{.........})}{100} = \text{.................}$$

C Practice questions

2 The table below shows the relative abundance of the two main isotopes of copper. Calculate the relative atomic mass of copper to one decimal place.

Isotope	^{63}Cu	^{65}Cu
Percentage abundance/%	69	31

3 The isotopes of sulfur and their abundance are shown in the table. Calculate the relative atomic mass of sulfur to two decimal places.

Isotope	Percentage abundance/%
^{32}S	95.02
^{33}S	0.76
^{34}S	4.22

The other kinds of average, less frequently used in GCSE Science, are mode and median. Mode means 'most common'. Median means 'in the middle', when arranged in increasing order.

Constructing frequency tables, bar charts and histograms

Data can be represented in a number of different ways. This section covers the use of frequency tables, bar charts and histograms.

Frequency tables

Scientists often have to collect data before it can be processed. One form of processing this data is in a frequency table. Frequency tables show the frequency (how many times something occurs) within a data set. Frequency tables are particularly useful for giving an overview of data, or for calculating the mode. There are two types of frequency table – ungrouped and grouped.

When you want to measure frequency, but no additional order is needed, you use an ungrouped frequency table. Suppose, for example, the lengths of 100 similar screws are measured to the nearest mm and their results are recorded in a table. This is an ungrouped frequency table.

Sometimes, there is so much data that it needs to be grouped into classes. For example, a physicist may be interested in the number of times the electronic components in certain pieces of apparatus fail, but it might be enough to know whether the number of failures was less than 5, between 5 and 10, between 10 and 15 and so on. This would require a grouped frequency table.

Both types of table are used to obtain statistical information, such as the mean, mode and the median (see pages 30-32).

Constructing frequency tables

When constructing frequency tables, you need to draw a table with the key data in the left-hand column and a tally column next to it.

In the following example, each time the diameter of a ball bearing is measured, a tally mark is placed in the appropriate part of the table. The first four tally marks are vertical, but the fifth tally mark is a diagonal across the previous four, so that the tallies are bundled into groups of 5. This makes it easier to count quickly – you can total the frequency in the right-hand column. Table 1.10 is an example of an ungrouped frequency table.

Table 1.10 **Example of an ungrouped frequency table**

Diameter, *D* (mm)	Tally	Frequency
21	IIII IIII I	11
22	~~IIII~~ II	7
23	~~IIII~~	4
24	IIII III	8
	Total	**30**

In Table 1.11, there are four classes for the masses of the ball bearings. This is a grouped frequency table.

Table 1.11 Example of a grouped frequency table

Mass, m (g)	Tally	Frequency
$10 \leqslant D < 15$	𝍸 𝍸 𝍸 I	16
$15 \leqslant D < 20$	𝍸 𝍸 𝍸 𝍸 II	22
$20 \leqslant D < 25$	𝍸 𝍸 IIII	14
$25 \leqslant D < 30$	𝍸 III	8
	Total	**60**

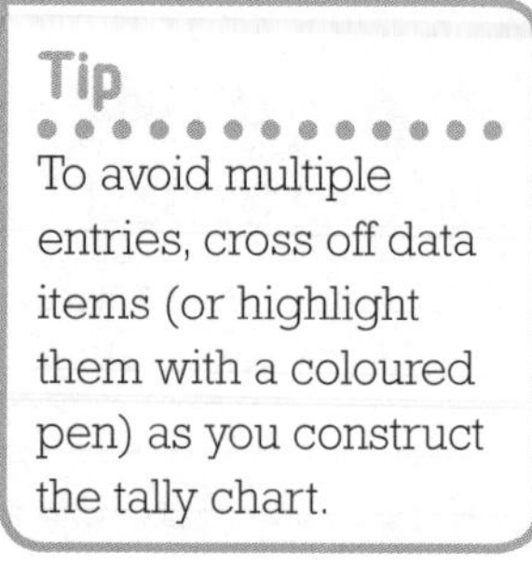

Tip

To avoid multiple entries, cross off data items (or highlight them with a coloured pen) as you construct the tally chart.

A Worked example

A Physics student measures the horizontal range *R*, in cm, of 40 marbles when they are rolled horizontally over the edge of a bench. The results are shown below.

Using groups of 0–5 cm, 5–10 cm, and so on, construct a grouped frequency table by means of a tally chart.

③	④	⑯	⑰	⑮	⑤	⑨	⑩	⑲	⑮
①	⑦	⑯	⑰	⑫	⑥	⑦	⑪	⑰	⑩
⑲	①	⑭	⑩	⑦	④	①	⑪	⑬	⑯
⑦	⑧	①	⑪	⑰	⑰	⑨	②	⑩	②

Range, *R* (cm)	Tally	Frequency
$0 \leqslant R < 5$	𝍸 IIII	9
$5 \leqslant R < 10$	𝍸 IIII	9
$10 \leqslant R < 15$	𝍸 𝍸	10
$15 \leqslant R < 20$	𝍸 𝍸 II	12
	Total	**40**

C Practice question

1 A Physics student wants to use dice to simulate radioactive decay. For the experiment to be valid, the dice must not be biased. This means that each one of the six numbers must be equally likely to turn up when the dice are thrown.

The student throws a single die 60 times and obtains the results shown below. The student is told that the die is probably unsuitable for the experiment if any number comes up fewer than eight times, or more than 12 times.

3	5	1	6	1	1	5	5	5	6	1	5
2	2	6	2	4	4	6	4	3	3	6	5
3	2	4	6	1	3	4	3	6	2	6	4
1	2	4	4	1	4	3	3	4	2	1	6
1	3	2	4	2	3	6	5	4	3	5	3

a Construct an ungrouped frequency table to represent the data.
b Use your table to decide whether or not the die is biased.

Using ungrouped frequency tables to find mean, mode and median

You can also use ungrouped frequency tables to quickly find mean, mode and median. For example, suppose the lengths of 100 similar engine components are measured to the nearest mm, and their results are recorded in a table. This is an ungrouped frequency table.

Length (mm)	98	99	100	101	102	**Total**
Frequency (number of items)	4	15	66	12	3	100

From this table we can calculate the mean, mode and median as follows.

Mean

We could set out the sum as 98 + 98 + 98 + 98 + 99 + 99 + 99 + (and so on).

However, there is a much quicker way. Multiply the frequency by the length, and add the totals together.

$$\text{mean length} = \frac{(4 \times 98) + (15 \times 99) + (66 \times 100) + (12 \times 101) + (3 \times 102)}{100} = 99.95\,\text{mm}$$

Mode

The number 66 is the highest frequency in the table. This shows there are 66 lengths of 100 mm. The mode is, therefore, 100 mm.

Median

Work out which number is in the middle when the items are arranged in order. In this case, there are 100 items and they are already in order of length. Since the number of items is even, there are two values in the middle. These are the 50th and the 51st values. The first 19 items are accounted for by the two lowest lengths. The next table entry shows 66 items each of length 100. So, the 50th and 51st items are to be found here. The median length is, therefore, 100 mm.

Using grouped frequency tables to find mean, mode and median

Grouped frequency tables can be used in a similar way. For example, suppose a physicist measures the mass of 40 different ball bearings (to the nearest gram) and presents her results as a grouped frequency table.

Class (g)	30–34	35–39	40–44	45–49	50–54	**Total**
Frequency	4	10	15	6	5	**40**

Mean

To calculate an estimate of the mean from this data, we imagine that, on average, the mass of each item in the class is the mid-point value (sometimes called the *class mark*). We can, therefore, add another row to the table.

Class (g)	30–34	35–39	40–44	45–49	50–54	**Total**
Frequency	4	10	15	6	5	**40**
Class mark (g)	32	37	42	47	52	

We can use the class mark (mid-point mass) to estimate the mean as we did with the ungrouped data.

$$\text{estimate of mean mass} = \frac{(4 \times 32) + (10 \times 37) + (15 \times 42) + (6 \times 47) + (5 \times 52)}{40}$$

$$= 41.75\,\text{g}$$

Note: This is only an *estimate* of the mean because it was assumed that all the items in any of the classes have the same value. For example, we have assumed that all four items in the class 30–34 have the value of 32.

Mode

The highest frequency in the table is 15, so the modal class is 40–44. The modal length within that class cannot be determined by methods available to GCSE students.

> **Tip**
> Categorical data is an example of discontinuous data.

Median

Since there are 40 items, we need to find the 20th and 21st items to find the median. There are 14 items with mass between 30 and 39 grams. The next table entry shows 15 items with masses between 40 and 44 grams. So, the 20th and 21st items are both within the class 40–44. The median class is, therefore, 40–44 grams.

> **Key terms**
>
> Bar charts: Charts showing discrete data in which the height of the unconnected bars represents the frequency.
>
> Discrete data: Data that can only have particular values, such as the number of marbles in a jar.
>
> Categorical data: Data that can take one of a limited number of values (or categories). Categorical data is a type of discontinuous data.
>
> Discontinuous data: Data that can have a limited range of different values, for example eye colour.
>
> Continuous data: Data that can have any value on a continuous scale, for example length in metres.

Bar charts

Bar charts are simple graphs used to show simple data. They can be used to show frequencies of categorical data (data that can be put into categories). The categories used are normally plotted on the x-axis, while frequency is on the y-axis. Unlike a histogram, the bars on a bar chart do not touch each other, to represent the fact that the bars represent distinct categories.

For example, a Physics student carried out a survey into the main type of heating used in 100 different households and recorded the results in the table below.

Type of central heating	Coal	Oil	Gas	Wind	Geothermal	Total
Number of households	4	15	66	12	3	100

We can use this data to draw a bar chart.

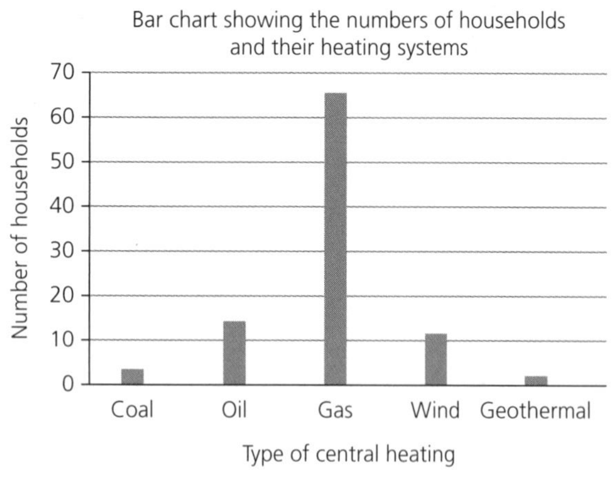

▲ Figure 1.5

> **Tip**
> This data is categorical because each type of heating is a distinct category and there is no overlap between the different categories. The best way to present such data is in a bar chart.

Each bar tells us the number of households using a particular type of heating. The longer the bar, the greater the number of households. Notice also:

- the axes are labelled exactly as in the table
- the bars are not joined together
- each bar has a label at the bottom.

Bar charts can also be used for continuous data.

> **Tip**
> Remember that the bars in a bar chart are never joined together. This is different to a histogram.

A Worked example

The table shows the pH of different household substances. Draw a bar chart of this data.

Household substance	pH
shampoo	6
mouthwash	10
baking soda	9
lemon juice	3
vinegar	3
water	7

Step 1 Decide what information should go on each axis and choose a suitable scale.

The horizontal x-axis should show the different types of household substance. The y-axis should show the pH.

Step 2 For the first substance – shampoo – draw a bar extending from the x-axis up to the correct value on the y-axis. Then leave a uniform space and draw the next bar, of the same width, for mouthwash.

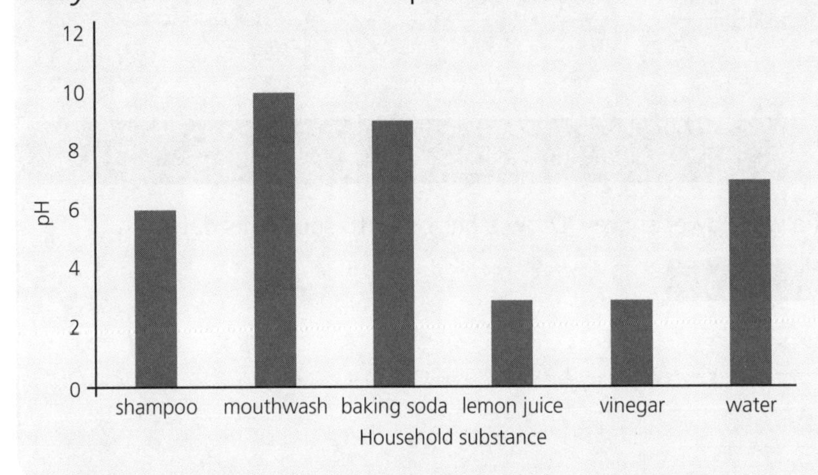

B Guided question

1 **Every year since 2014, the Physics department in a sixth form college has recorded the numbers of male and female students taking GCSE Physics. The results from 2018 to 2014 are shown in the bar chart.**

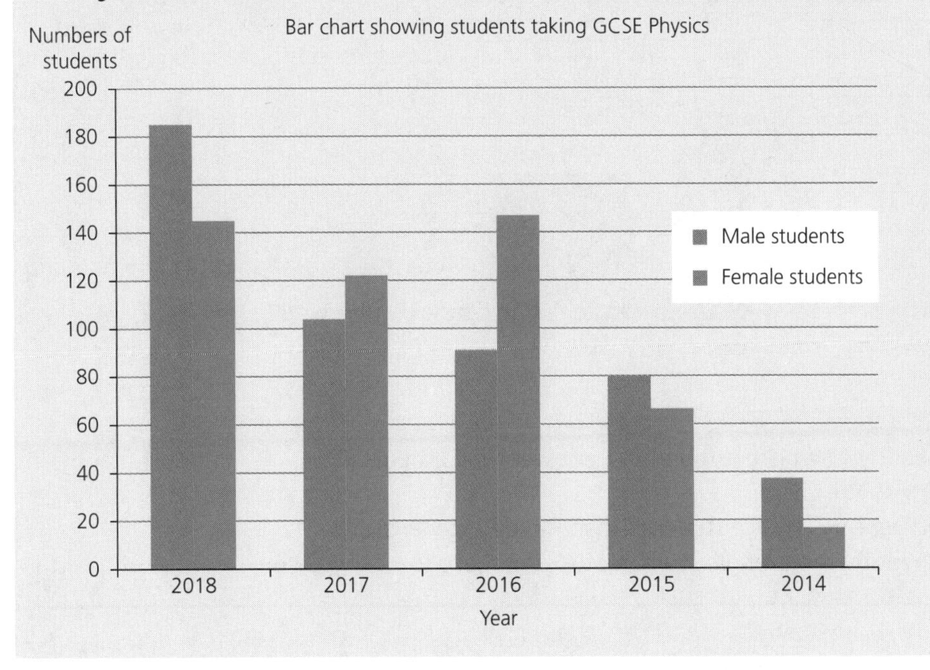

> **Tip**
> This question shows how you can also place two bars side-by-side.

Use the chart to find:

a **the year in which the number of female students was greatest**

Step 1 Female students have a ....-coloured bar.

Step 2 This bar is biggest in ....

b **the year in which the difference between the numbers of male students and female students was greatest**

Step 1 Year in which difference in males and females is greatest is the year in which there is the greatest difference in the .... of the bars.

Step 2 This year is ....

c **the total number of students taking GCSE Physics in 2015.**

Step 1 Number of male students in 2015 = ...

Step 2 Number of female students in 2015 = ...

Step 3 Total number of students in 2015 = ... + ...
= ....

Tip

Scales can start at zero, but do not have to.

C Practice questions

2 The table below shows the results of a wildflower survey. Draw a bar chart to show this data.

Wildflower species	Frequency
Cow parsley	16
Daisy	52
Cornflower	4
Cowslip	23

3 Below is an incomplete bar chart of a survey carried out to find the average time spent by students studying Biology, Chemistry, Mathematics and Physics every week in a school.

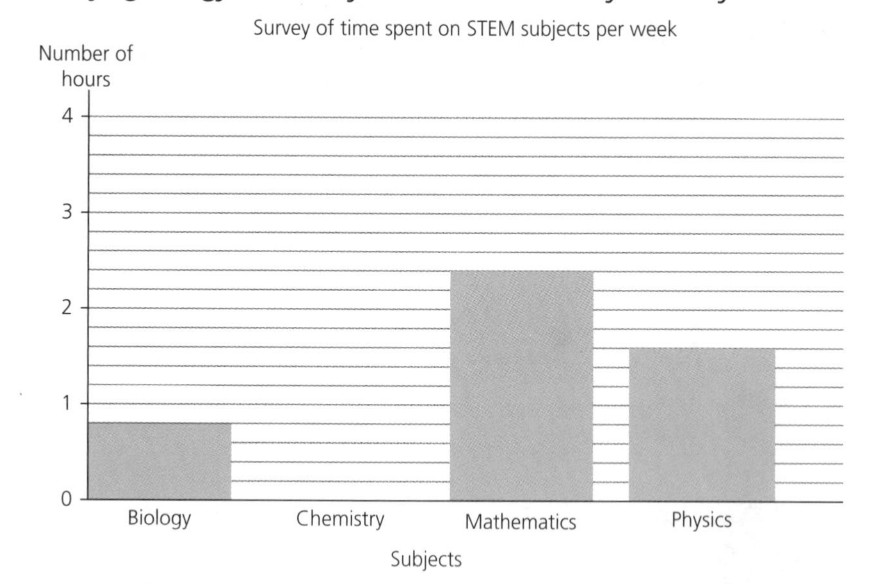

a The number of hours spent by Chemistry students every week is 3.3 hours. Add the bar for Chemistry to the chart.
b How much more time is spent by Physics students than Biology students?
c What is the total time spent by the average student studying these four subjects?

Histograms

Histograms are similar to bar charts but are only used for continuous data, such as the length or mass of components. This means that in a histogram the bars *must* touch.

A Worked example

A physicist measured the wavelengths, to the nearest 10 cm, at which empty lemonade bottles resonated. She recorded the results in a table, as shown.

Show these data in a histogram.

Wavelength λ (cm)	150	160	170	180	190	200	210	220	230
Number of bottles	2	3	4	5	6	7	8	6	5

Step 1 Since the wavelengths are measured to the nearest 10 cm, the 150 cm wavelength covers 145 cm $\leq \lambda <$ 155 cm, the 160 cm wavelength covers 155 cm $\leq \lambda <$ 165 cm, and so on.

Step 2 These limits are the lower and upper boundaries of the wavelengths. They allow us to draw this grouped frequency table.

Wavelength λ (cm)	Class limits	Number of bottles
150	145 cm $\leq \lambda <$ 155 cm	2
160	155 cm $\leq \lambda <$ 165 cm	3
170	165 cm $\leq \lambda <$ 175 cm	4
180	175 cm $\leq \lambda <$ 185 cm	5
190	185 cm $\leq \lambda <$ 195 cm	6
200	195 cm $\leq \lambda <$ 205 cm	7
210	205 cm $\leq \lambda <$ 215 cm	8
220	215 cm $\leq \lambda <$ 225 cm	6
230	225 cm $\leq \lambda <$ 235 cm	5

Step 3 In this case, the width of each bar is always 10 cm as can be seen from the class limits.

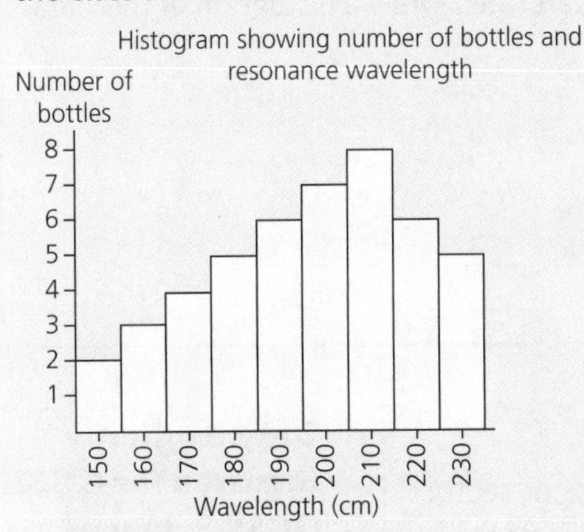

Step 4 The wavelength corresponding to 150 cm is a bar stretching from 145 cm to 155 cm, which are the class limits. The adjacent bar stretches from 155 to 165 cm, and so on. In this way the bars touch – because the data is continuous.

Key term

Histograms: Charts showing continuous data in which the area of the bar represents the frequency.

Tip

The histograms shown in this book are all of equal width as histograms with bars of unequal width are beyond the scope of GCSE Science.

B Guided question

1 **The amount of snow that fell over a 20 day period in a winter resort was recorded. The results are shown in this table.**

Snow fall S (mm) (class)	Class mid-point snow fall (mm)	Number of days
$10 \leqslant S < 20$	15	3
$20 \leqslant S < 30$		6
$30 \leqslant S < 40$		
$40 \leqslant S < 50$		3
$50 \leqslant S < 60$		2

Show this data on a histogram.

Step 1 The number of days have to add up to 20.

So, the number of days when the snowfall was between 30 and 40 mm was ....

Step 2 The class mid-point is exactly half way between the maximum and minimum snowfall.

So, the numbers missing from the middle column of the table are ...., ..., ..., and ....

Step 3 Draw a set of axes. The vertical axis is labelled *Number of days*. The horizontal axis is labelled .... and will range from 0 to ....

Step 4 The first bar is centred on 15 mm, ... days in height and .... mm wide. Draw this bar.

Step 5 Draw the remaining bars. The final bar is two days in height, centred on .... mm and .... mm wide.

Step 6 Add the title to the histogram.

C Practice question

2 The table below shows the results of an investigation into resting heart rates. Draw a histogram of this data.

Resting heart rate (bpm)	Frequency
60–69	14
70–79	59
80–89	132
90–99	97

Pie charts

★ **Only explicitly required for CCEA GCSE Science Single and Double Award.**

Pie charts display the proportions of a whole data set as sector angles or sector areas. The total angle around the centre of the pie is 360° and the area of the sector is proportional to the angle at the centre.

Suppose a Physics student investigated the way in which 60 different students came to school. The data collected was to be shown in a pie chart. We can use this information to work out the following:

60 students = 360°

So, 1 student = 360 ÷ 60 = 6°

This means that, for each sector, the angle at the centre of the pie chart would be calculated using:

angle = number of students $\times 6^\circ$

A Worked example

This pie chart represents the energy resources of a European country.

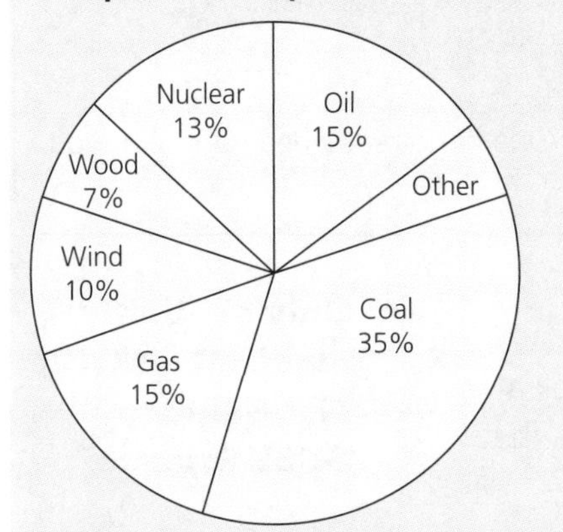

a What percentage of the energy resource is represented by 'Other'?

Step 1 Total percentages given = 35% + 15% + 10% + 7% + 13% + 15% = 95%

Step 2 So, 'Other' represents 100% − 95% = 5%

b What is the sector angle of the Wind sector?

Step 1 The Wind sector represents 10% of the pie.

Step 2 So, its sector angle is 10% of $360^\circ = \frac{10}{100} \times 360 = 36^\circ$

c The sector labelled Gas represents an energy of 8.1×10^{17} J.

Calculate the total energy resources used by this country.

Step 1 Gas represents $15\% = 8.1 \times 10^{17}$ J

Step 2 $1\% = \frac{8.1 \times 10^{17}}{15} = 5.4 \times 10^{16}$

Step 3 $100\% = 100 \times (5.4 \times 10^{16}) = 5.4 \times 10^{18}$ J

B Guided question

1 The results of a survey into the method used by pupils to come to school showed the following:

Method	Walking	Cycling	Car	Bus	Train
Number of pupils	15	5	35	20	15

Show the results in a pie chart.

Step 1 The number of students surveyed altogether was: ....

Step 2 Each student in the pie chart is represented by an angle of .... degrees.

Step 3 So, the angles for each method of transport are:

Walking = 60°; Cycling = ... °; Car = ... °; Bus = ... °; Train = ... °.

Step 4 With a compass, draw a large circle to represent the pie.

Step 5 With a ruler, draw a line from the centre of the circle to its circumference.

Step 6 Draw the sectors in the pie using the angles found in Step 3, then label the sectors.

C Practice question

2 Below are the results of a survey of 180 people asked about the main energy resource used to heat their homes.

Energy resource	Gas	Oil	Coal	Electricity	Wood	Other
Number of people	90	45	25	10		2
Sector angle (degrees)			50			

a Calculate the number of people whose main energy resource is wood.
b Calculate the sector angle for each resource, if displayed as a pie chart. One has already been done for you.
c Show the data on a pie chart.

Understanding the principles of sampling

★ **Only explicitly required for OCR 21st Century GCSE Science.**

Sampling is a very wide-ranging practical topic. In this section, we will look at the mathematics involved in some of the ecological sampling techniques that may feature in GCSE Science.

Quadrats

Quadrats are tools for assessing the abundance of non-mobile organisms such as plants. Quadrats can be used to estimate:

- species frequency: the number of individuals of a certain species found in the sample area
- species density: the number of individuals of a certain species per unit area
- percentage cover: the percentage of the quadrat area that is occupied by individuals of a particular species. This measure is particularly useful for species where it would be difficult to count every individual plant, for example grasses.

Key term

Ecological: The relation of living organisms to one another and to their physical surroundings.

Mark and recapture

This sampling technique allows us to estimate the number of mobile organisms (for example woodlice) in a particular area.

First, a number of organisms of a certain species are caught from a defined area and marked. These organisms are then released. After a period of time, the same area is sampled again, and the marked individuals among this second sample are counted. The total population size can then be estimated using the equation:

Population size = (total number in first sample × total number in second sample) ÷ number marked in second sample

A Worked examples

1 **In a sampling activity, 0.25 m^2 quadrats were placed randomly in a 10 m by 10 m grid. Each quadrat was divided into 25 equal squares. In quadrat 1, grass filled 20 of the squares in the quadrat. What was the percentage cover of grass in this quadrat?**

To find the percentage cover:

Step 1 Divide the area of the quadrat that is covered by the organism by the total area of the quadrat: $20 \div 25 = 0.80$

Step 2 Multiply this answer by 100: $0.80 \times 100 = 80\%$

So the percentage cover of grass in this quadrat is 80%.

B Guided questions

1 **During a survey of a crab population, 92 crabs were caught, tagged and then released. Several months later, 78 crabs were caught and of these, 15 had been tagged. Use the equation below to calculate the size of the crab population.**

Population size = (total number in first sample × total number in second sample) ÷ number marked in second sample

Step 1 Number in first sample = 92

Number in second sample = 78

Number marked in second sample = 15

Step 2 population size =

(.................. × ....) ÷ number marked in second sample

Step 3 Population size = ...

2 **A survey of *Digitalis* was carried out in a field. The results of 10 quadrats are shown below. The area of each quadrat is 0.25 m^2.**

Quadrat	1	2	3	4	5	6	7	8	9	10
Number of *Digitalis*	2	1	3	0	0	2	1	3	0	4

What was the species frequency of *Digitalis*?

Step 1 Count how many of the quadrats contained *Digitalis*.
The number is: ....

Step 2 Divide the number of quadrats containing *Digitalis* by the total number of quadrats, and multiply by 100 to get a percentage:

Population size = (.................. ÷ ....) × 100 = ...%

Tip

In an exam question asking you to do calculations about capture-recapture, you will be given the equation.

C Practice questions

3 In an investigation into the population of snails in an area, 105 were caught and marked. Two weeks later, the sampling activity was repeated. Out of 120 snails caught, 45 were marked. Estimate the total size of the snail population.

4 During an investigation into a population of a species of grass, a quadrat with 25 equal squares was used to estimate the percentage cover. In one of the quadrats, 15 of the squares were covered with the grass. Estimate the percentage cover of the grass in this quadrat.

Simple probability

★ **Only explicitly required for OCR 21st Century Combined Science, WJEC/Eduqas and Edexcel International GCSE Science Double Award.**

Probabilities are usually expressed as decimals or fractions, and sometimes as percentages. In GCSE Biology, you will encounter probability in the context of genetic crosses, whilst in Physics it will likely crop up in the context of radioactive decay. If something is certain to happen, it has a probability of 1 (or 100%). If something is certain not to happen, it has a probability of 0. All probabilities must lie between 0 and 1.

The sum of the probabilities of *all* possible outcomes of an experiment is 1. Therefore, if the probability of an event occurring is 0.25, then the probability of that event not occurring is 0.75 (because 0.25+0.75=1).

The probability of an event, E, occurring is defined as the ratio:

$$\text{probability of E occurring} = \frac{\text{total number of favourable outcomes}}{\text{total number of possible outcomes}}$$

Suppose that the probability that any individual nucleus will decay in a given minute is 0.2, and that initially we have a population of 1000 undecayed nuclei.

After one minute, 200 would be expected to decay, leaving 800 undecayed.

After a further minute, we would expect 160 more to decay, leaving 640 nuclei undecayed.

This process would continue as shown:

Time elapsed (mins)	0	1	2	3	4	5	6	7
Expected number of undecayed nuclei	1000	800	640	512	410	328	262	210

A quick look at the table shows that the time taken for the number of undecayed nuclei to fall to half of its original number is a little over 3 minutes. Scientists recognise this as the half-life for that decay.

Tip

Phrases such as 'probability of E occurring' are very wordy. Physicists tend to shorten it to P(E).

A Worked examples

1 **A species of snake can either have red or yellow markings. In a particular population of snakes, the probability of a snake having red markings is 0.65. What is the probability of a snake in this population having yellow markings?**

We know that:

- red or yellow are the only possible marking colours
- the probabilities of all possible outcomes must add up to 1
- probability of red markings + probability of yellow markings = 1
- rearranging this equation gives:
  probability of yellow markings = 1 – probability of red markings
  = 1 – 0.65 = 0.35
- so the probability of a snake in this population having yellow markings is 0.35.

2 **In humans, the sex chromosomes (X and Y) determine biological gender. Males have XY and females have XX. Use a genetic cross to show the probability of a couple having a baby girl. Give your answer as a percentage.**

Step 1 Write out the genotypes of the parents.

Parents: XX XY

Step 2 Write out the gametes produced by the parents.

Gametes: X X X Y

Step 3 Use the gametes to draw a Punnett square to find the genotype of the offspring.

	X	Y
X	XX	XY
X	XX	XY

Expected offspring ratio = 2 female (XX) : 2 males (XY) = 1 female : 1 male

Tip

As probabilities are concerned with chance, there is always a possibility that the observed results do not match the expected results. This is the reason that, despite the most likely outcome of having two children being a boy and a girl, many parents actually have two girls or two boys.

Step 5 To calculate the probability of having a girl from this ratio, divide the number of females by the total of the numbers in the ratio. As the question asked for the answer as a percentage, you have to multiply your answer by 100

$$\frac{1}{2} \times 100 = 50\%$$

There is a 50% chance of the couple having a girl.

B Guided question

1 **This graph shows a decay curve for a radioisotope.**

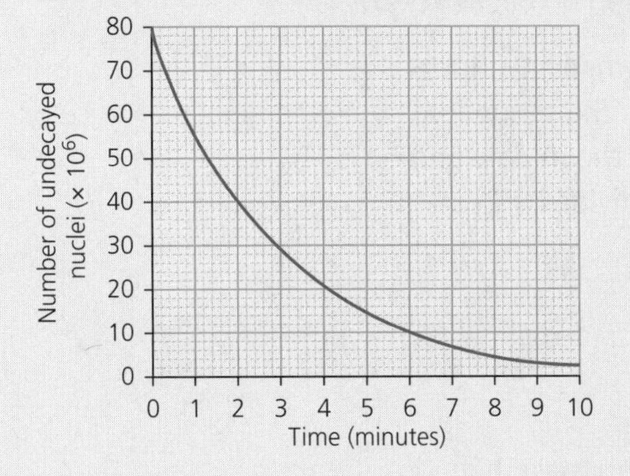

a **Find the half-life of the isotope starting with the original number of undecayed nuclei.**

Step 1 In one half-life, the number of undecayed nuclei falls by .... %

Step 2 So, in one half-life the 80 million undecayed nuclei will fall to .... million.

Step 3 From the graph this takes .... minutes.

b **Use your answer to part a to state the probability that a particular nucleus in the sample is likely to decay within a period of 6 minutes.**

Step 1 In 6 minutes, the number of undecayed nuclei has fallen to .... million.

Step 2 So, the fraction that has decayed is ....

Step 3 So, the probability of decay within 6 minutes is ....

Tip

Although we might know the probability that a particular nucleus will decay in a given time interval, we cannot say for certain whether or not it will decay. The process is random and spontaneous. This means that the experimental graphs for decay are likely to be more ragged than those you see in textbooks or examination papers.

C Practice questions

2 Suppose a sample contains 3000 undecayed nuclei and that the probability that a particular nucleus will decay in a given minute is 0.3.

a Copy and complete the table below to show the expected number of undecayed nuclei every minute up to 7 minutes. Two entries have been completed for you.

Time elapsed (mins)	0	1	2	3	4	5	6	7
Expected number of undecayed nuclei	3000					504		

b Plot the graph of *Expected number of undecayed nuclei* (y-axis) against *Time elapsed in minutes* (x-axis).

c Use your graph to show that the half-life of this decay is approximately 1.9 minutes.

3 In a species of plant, green fruit is the dominant phenotype and yellow fruit is the recessive phenotype. A genetic diagram was used to estimate the results of crossing two heterozygous plants.

Use a Punnett square diagram to determine the probability of the offspring having yellow fruit.

Use the following symbols: G = dominant allele; g = recessive allele

Understanding mean, mode and median

The three types of average that you will come across in your GCSE exam questions are:

- mean: this is the average, and is covered in the 'means' section on pages 26-28
- median: this is the middle value in the data set. To find the median, arrange all the data points in order and pick out the middle value in the sequence. If there are an even number of data points, take the two in the middle and calculate their mean (for example add them together and divide by 2)
- mode: this is the most common value in the data set.

The most appropriate type of average to use depends on the context:

- The most commonly used average is the mean.
- The median is more useful than the mean if there are exceptionally high or low values (outliers) in the data, which would skew the mean.
- The mode is suitable for use with non-numerical data or when the data points cannot be put in a linear order.

Suppose a student measures, to the nearest gram, the mass of 21 marbles and sets out the results in order of increasing mass, as below.

14, 14, 14, 15, 15, 15, 15, 16, 16, 16, **16**, 17, 17, 17, 17, 17, 17, 17, 17, 18, 18

In this example, the most common number in the list is 17. This means that the modal mass is 17 grams (or simply the mode is 17 g).

Now look at the highlighted number, **16**. Of the 21 numbers in the list, 10 are to the left of it and 10 are to the right. The number 16 is in the middle. So, the median mass is 16 grams.

It's easy to find the number in the middle when the total number is odd. Just add 1 to the total number of results (do not add all the results together) and divide by 2. In this case, $(21 + 1) \div 2 = 11$, so the median is the 11th number counting from either end.

It's slightly trickier when there is an even number of numbers, as there are two numbers in the middle. Consider the following two lists of six numbers.

List A: 2, 3, **5**, **5**, 6, 7   List B: 2, 3, **4**, **5**, 6, 7

In List A, the middle numbers are both 5. So, the median is 5.

In List B, the middle numbers are 4 and 5. They are different, and so we take the median to be half-way between them. So, the median is 4.5.

A Worked example

The following table shows the raw results of a survey of blood groups.

A	AB	AB	O	O	O
B	AB	A	A	B	AB

Find the mode of these blood groups.

Step 1 Draw a frequency table of each value.

Blood group	Frequency
A	3
B	2
O	3
AB	4

Step 2 Determine which is the most common blood group. AB has a frequency of 4, so the mode blood group is AB.

B Guided question

1 **The diameters of nine ball bearings are measured to the nearest mm. The mode is 9 mm and the median is 6 mm. Seven of the nine diameters, in mm, are:**

4, 1, 7, 6, 2, 3, 9

The largest diameter is 9 mm.

Find the missing diameters if both of them are known to be greater than 6 mm.

Step 1 Arrange the numbers in order: ...., ..., ..., ..., ..., ..., ...,

Step 2 Since the mode is 9, there must be at least .... 9s. So, add .... to the ordered list. There are now .... numbers in the ordered list.

Step 3 The median is the .... number in the ordered list, so the missing number must be ...., ..., ... or .... Since the mode is 9, the missing number cannot be .... or .... So, the missing number must be .... or ....

C Practice question

2 Twenty students carry out an experiment to find the density of ethanol. Below are their results, in g/cm^3.

0.79	0.79	0.78	0.78	0.78	0.77	0.77	0.79	0.79	0.77
0.77	0.79	0.79	0.79	0.79	0.78	0.78	0.78	0.78	0.77

Find:

a the mode

b the median density, in g/cm^3.

Using a scatter diagram to identify a correlation

When data points are plotted on a scatter graph, it may be possible to identify correlations in the data. A correlation can be either positive or negative.

★ **Not specifically required by WJEC/Eduqas or CCEA GCSE Science Single or Double Awards.**

A positive correlation is one where as one variable increases, the other variable also tends to increase:

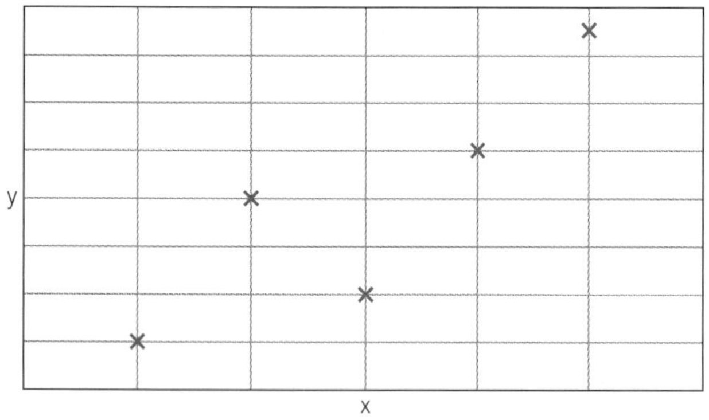

▲ Figure 1.6 Scatter diagram showing a positive correlation

A negative correlation means that as one variable increases, the other variable shows a decreasing trend.

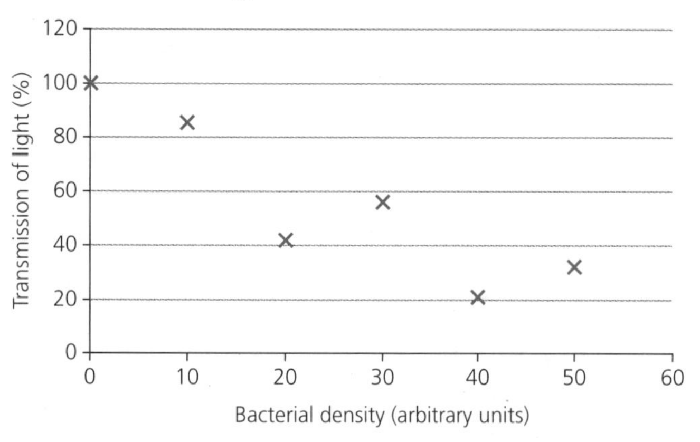

▲ Figure 1.7 Scatter diagram showing a negative correlation

In some situations it will be clear that correlation exists, but occasionally some data will show no correlation. If you are asked about scatter graphs in the exam, any correlation (positive, negative or none) should be evident.

Two points to note:

① Do not think that, because there is positive correlation, there is a causal relationship. If you were to investigate the number of trees growing in a person's garden, you might find a correlation with the number on their front door. But that is not to say the number of trees increases just because the number on the door increases.

② On the other hand, a positive correlation *might* also show a causal relationship. For many years, tobacco companies said that there was a correlation between smoking and lung cancer, but there was no causal relationship and it was just coincidence. Now on every pack of cigarettes there is a clear message: Smoking Kills.

Tip

To identify correlation, ask yourself what happens to one variable when the other increases.

- If one goes up when the other goes up, it is positive correlation.
- If one goes up when the other goes down, it is negative correlation.
- If there is no relationship, it is no correlation.

Key terms

Scatter graph: A graph plotted between two quantities to see if there might be a relationship between them.

Positive correlation: This occurs if one quantity tends to increase when the other quantity increases.

Negative correlation: This occurs if one quantity tends to decrease when the other quantity increases.

No correlation: There is no relationship whatever between two quantities.

Causal relationship: The reason why one quantity is increasing (or decreasing) is that the other quantity is also increasing (or decreasing).

A Worked example

The scatter diagram below shows the effect of dissolved oxygen concentration on the population of rainbow trout in an aquaculture pond. What type of correlation is shown by this data?

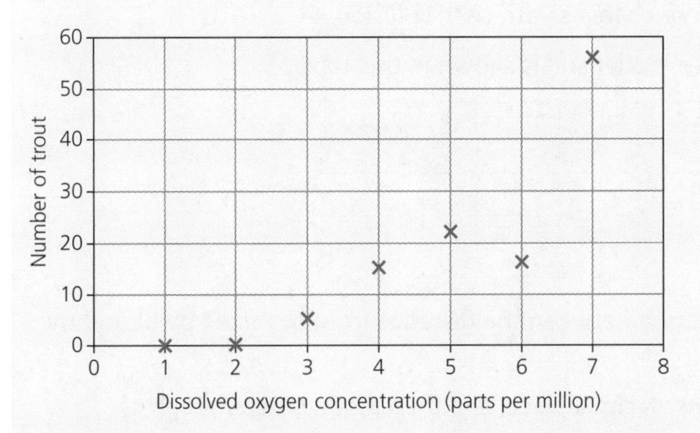

Step 1 Look for a general trend in the distribution of the points on the scatter diagram. As you move from left to right in the diagram, do the data points seem to be getting higher or lower?

Step 2 As the dissolved oxygen concentration increases, the population of the rainbow trout also increases. Not every data point fits the pattern exactly, but this should not be surprising, as data involving living organisms will always show variation.

Step 3 State the correlation shown. The scatter diagram indicates a positive correlation between the dissolved oxygen concentration and the population of rainbow trout.

B Guided question

1 **The scatter graph below shows how the percentage cover of a grass species varies with distance from a large tree. What type of correlation is shown by this data?**

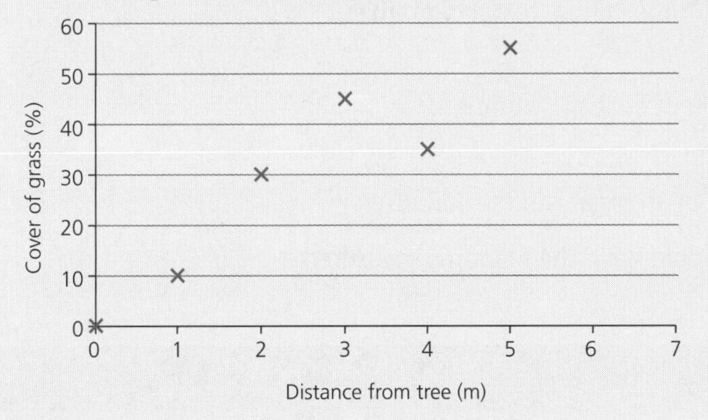

Step 1 Describe the correlation shown by the graph.

As the distance from the tree increases, the % cover of grass increases.

Step 2 This shows a .... correlation.

C Practice question

2 In a computer model of the molecules of a gas, each molecule changes its direction randomly when it collides with another molecule. The time between the collisions is also random. The distance a particular molecule is from a fixed point therefore changes with each collision.

A small data set is generated from the computer model and is shown in this table.

Distance D of molecule from a fixed point (mm)	0	15	21	26	30	33	38	40
Collision number N	0	1	2	3	4	5	6	7
$\sqrt{N}$	0		1.4		2.0			

It has been suggested that there is some correlation between the distance from the fixed point and the square root of the collision number N.

Complete the table, entering the numbers to one decimal place, and then plot a scatter graph of *Distance D* against $\sqrt{N}$.

What type of correlation, if any, is suggested by the scatter graph?

Orders of magnitude

★ **Not explicitly required for CCEA GCSE Science Single or Double Award.**

Orders of magnitude allow us to compare very large and very small values to each other. This is useful when comparing the size of different types of particles, interconverting units and using the magnification equation.

An order of magnitude is a division or multiplication by 10. Each division or multiplication by 10 is termed an order of magnitude. The actual length may be approximated as it is the relative difference that is important. Orders of magnitude are easily compared using standard form (see page 9).

Prefixes are used to change the magnitude of a unit. The same prefixes are used for all units. See Table 1.3 on page 2 to remind yourself how to apply prefixes correctly to units.

It is important that you select the appropriate unit for each situation, for example:

- It would be inappropriate to give the length of an organism in kilometres.
- Only the smallest organisms or structures would have their lengths measured in micrometres.

A Worked example

The radius of an atom is 1×10^{-10} m. The radius of the nucleus of an atom is 1×10^{-14} m. How many times bigger is the atom than the nucleus?

You can compare the two diameters by dividing the larger power of 10 by the smaller one.

$$1 \times 10^{-10} \div 1 \times 10^{-14} = 10^{4}$$

The diameter of the atom is 10000 (10^4) times bigger than that of the nucleus.

B Guided question

1 **In the hydrogen atom an electron orbits a proton. The gravitational force on the electron is 4.06×10^{-47} N. The electrical force on the proton is 9.22×10^{-8} N.**

Calculate how many times greater the electrical force is than the gravitational force.

Give your answer as an order of magnitude.

Step 1 Electrical force ÷ gravitational force
= .... N ÷ .... N = ...

Step 2 So, as an order of magnitude, the electrical force is .... times more than the gravitational force.

Tip

When doing order of magnitude calculations, you often have to enter into your calculator numbers in standard form. It is worthwhile referring back to the section on page 11 to revise how to do that before attempting such questions.

C Practice questions

2 A coarse particle has a diameter of 1×10^{-4} m and a nanoparticle has a diameter of 1×10^{-9} m. Calculate how much bigger a coarse particle is than a nanoparticle.

3 A fine particle has a diameter of 1.0×10^{-6} m. A nanoparticle has a diameter of 1.6×10^{-9} m. Calculate how many times bigger the diameter of the fine particle is than the diameter of the nanoparticle.

Using the magnification equation

Biology often involves studying organisms and structures that are incredibly small. The magnification equation makes it possible to determine the actual dimensions of these organisms and structures from micrographs and scale drawings. It also helps you to create your own scale drawings.

The magnification equation is:

Magnification = image size ÷ object size

You should be confident using and rearranging this equation to calculate:

- the magnification if you are given an image size and an object size
- the image size if you are given a magnification and an object size
- the object size if you are given a magnification and an image size.

An exam question on this skill might involve measuring part of a diagram or electron micrograph, so make sure you can do this accurately.

A Worked example

An electron micrograph of a nucleus showed it as having a diameter of 80 mm. The actual diameter of the nucleus was labelled as 0.0004 mm. What was the magnification of the electron micrograph?

Step 1 Identify the image size and the object size. The image size is the diameter of the electron micrograph. The object size is the actual diameter of the nucleus. Therefore: Image size = 80 mm; object size = 0.0004 mm

Step 2 Substitute the values into the magnification formula:

Magnification = image size ÷ object size

Magnification = 80 ÷ 0.0004

Magnification = 200 000

Hence the magnification is 200 000.

Tip

Magnification is actually a ratio and therefore does not have a unit.

B Guided question

1 A student used a microscope to make a drawing of a transverse section of a root. The diameter of the drawing was 150 mm. The student measured the width of the root using the microscope as 2 mm. Use the magnification equation to calculate the magnification of the student's drawing.

Step 1 Identify the image size and the object size.

Step 2 Substitute the values into the magnification formula and calculate:

Magnification = image size ÷ object size

Magnification = ... ÷ .... = ...

C Practice question

2 On a diagram of a fungal cell, the cell had a width of 30 mm. The magnification was given as 340 ×. What was the actual width of the fungal cell? Give your answer in standard form to 2 significant figures.

Algebra

Algebra is a branch of mathematics that uses equations in which letters represent numbers.

You need to know how to solve various equations by:

- rearranging the equation to change the subject – this step is only required if the letter (or value) on its own is not what you want to calculate
- substituting in the correct numbers for each letter or value
- calculating the answer.

Chemistry formulae

Table 1.12 lists some of the key Chemistry equations required in the GCSE exam board specifications. These may not be provided in the exam, so you may need to learn them, depending on which specification you are using. Ask your teacher if you're not sure:

Table 1.12 Key Chemistry equations

amount in moles $= \frac{\text{mass (g)}}{M_r}$ where M_r = relative formula mass
amount in moles $= \frac{\text{mass (g)}}{A_r}$ where A_r = relative atomic mass
percentage yield $= \frac{\text{actual yield}}{\text{theoretical yield}} \times 100$
atom economy $= \frac{\text{sum of relative formula mass of desired product from equation}}{\text{sum of relative masses of all reactants from equation}} \times 100$
amount in moles of a gas $= \frac{\text{volume (dm}^3\text{)}}{24}$

amount in moles $= \dfrac{\text{volume (cm}^3) \times \text{conc.(mol/dm}^3)}{1000}$
mean rate of reaction $= \dfrac{\text{quantity of reactant used}}{\text{time taken}}$
$R_f = \dfrac{\text{distance moved by substance}}{\text{distance moved by solvent}}$

Physics formulae

Table 1.13 lists the Physics equations in the specifications of the GCSE exam boards. Most of them you have to remember and be able to use, but use the guide below to check for your board's specifics.

Table 1.13 Equations in the specifications of the GCSE exam boards

KEY
RR = required recall (you must remember the equation and be able to use it).
S&U = select and use (you will be given a list of formulae and you need to be able to select the correct one and use it – but you do not need to remember it).
N/A = not required for this specification.
DS = formula given on data sheet in examination.
Highlight = the equation will only appear in Higher Tier papers.

Context	Equation	AQA	OCR	Edexcel UK	WJEC	CCEA	Edexcel International
Weight	W = mg	RR	RR	RR	RR	RR	RR
Work	W = Fs or E = Fd	RR	RR	RR	RR	RR	RR
Hooke's Law	F = ke or F = kx	RR	RR	RR	RR	RR	RR
Moment	M = Fd	RR	RR	RR	RR	RR	**S&U**
Pressure	$p = \dfrac{F}{A}$	RR	RR	RR	S&U	RR	RR
Distance	s = vt	RR	RR	RR	RR	RR	RR
Average speed	$\bar{v} = \dfrac{1}{2}(u + v)$	RR	N/A	N/A	RR	RR	RR
Acceleration	$a = \dfrac{\Delta v}{t}$	RR	RR	RR	RR	RR	RR
Newton's Law	F = ma	RR	RR	RR	RR	RR	RR
Momentum	p = mv	RR	**RR**	**RR**	RR	N/A	**S&U**
Kinetic energy	$E_k = \dfrac{1}{2}mv^2$	RR	DS	RR	RR	RR	RR
GPE	$E_p = mgh$	RR	RR	RR	RR	RR	RR
Power	$P = \dfrac{E}{t}$ or $P = \dfrac{W}{t}$	RR	RR	RR	RR	RR	RR
Efficiency	efficiency $= \dfrac{\text{useful output energy}}{\text{total input energy}}$	RR	RR	RR	S&U	RR	RR
	efficiency $= \dfrac{\text{useful output power}}{\text{total input power}}$	RR	N/A	N/A	S&U	RR	N/A
Wave equation	$v = f\lambda$	RR	RR	RR	S&U	RR	RR
Charge	Q = It	RR	RR	RR	N/A	RR	RR
Ohm's Law	V = IR	RR	RR	RR	S&U	RR	RR
Series resistance	$R_T = R_1 + R_2$	RR	RR	RR	S&U	RR	N/A

Context	Equation	AQA	OCR	Edexcel UK	WJEC	CCEA	Edexcel International
Parallel resistance	$\frac{1}{R_T} = \frac{1}{R_1} + \frac{1}{R_2}$	N/A	N/A	N/A	S&U	RR	RR
Joule's Law (power)	$P = VI$	RR	RR	RR	RR	RR	RR
Joule's Law (power)	$P = I^2R$	RR	RR	RR	RR	N/A	RR
Power	$E = Pt$ or $E = IVt$	RR	RR	DS	S&U	RR	RR
Voltage	$E = QV$	RR	RR	RR	N/A	RR	RR
Density	$\rho = \frac{m}{V}$	RR	RR	RR	S&U	RR	RR
Pressure	$P = h\rho g$	RR	DS	DS	RR	N/A	RR
Motion equation	$v^2 = u^2 + 2as$	RR	DS	DS	RR	N/A	RR
Force (due to an impulse)	$F = \frac{m\,\Delta v}{\Delta t}$	RR	N/A	DS	RR	N/A	S&U
Heat capacity	ΔE (or Q) $= mc\Delta\theta$	RR	DS	DS	S&U	N/A	S&U
Periodic time	$T = \frac{1}{f}$	RR	N/A	N/A	N/A	RR	RR
Magnification	$M = \frac{h_{image}}{h_{object}}$	RR	N/A	N/A	N/A	N/A	N/A
Force in magnetic field	$F = BIl$	RR	DS	DS	RR	N/A	N/A
Latent heat	$E = mL$ or $Q = mL$	RR	DS	DS	RR	N/A	N/A
Transformer turns-ratio	$\frac{V_p}{V_s} = \frac{n_p}{n_s}$	RR	DS	DS	RR	RR	S&U
Transformer power	$V_sI_s = V_pI_p$	RR	DS	DS	N/A	N/A	S&U
Boyle's Law	$pV = \text{constant}$ or $p_1V_1 = p_2V_2$	RR	DS	DS	RR	N/A	RR
Pressure Law	$\frac{p_1}{T_1} = \frac{p_2}{T_2}$	N/A	N/A	N/A	N/A	N/A	RR
Gas Law	$\frac{PV}{T} = \text{a constant}$	N/A	N/A	N/A	S&U	N/A	N/A
Energy stored in spring	$E = \frac{1}{2}kx^2$	DS	DS	DS	RR	N/A	RR
Snell's Law	$n = \frac{\sin i}{\sin r} = \frac{1}{\sin c}$	N/A	N/A	N/A	N/A	N/A	RR

Understanding and using algebraic symbols

The following symbols are often used in algebra, and you may encounter them in GCSE Science exam questions where equations are used. You should learn to recognise their meanings as given in Table 1.14.

Table 1.14 Symbols in algebra

Symbol	Meaning
$=$	equals
$>$	greater than
$\geqslant$	greater than or equal to
$<$	less than
$\leqslant$	less than or equal to
$\propto$	proportional to
$\sim$	approximately

Inequalities show a relation between two values that are not equal.

A Worked examples

1 **The table below shows the rate of photosynthesis and rate of respiration in a plant at different times of the day.**

Time	Rate of respiration / arbitrary units	Rate of photosynthesis / arbitrary units
8 a.m.	40	70
12 p.m.	70	100
10 p.m.	50	0

Write inequalities that relate the rate of respiration to the rate of photosynthesis at 12 p.m. and 10 p.m.

Step 1 Determine which of the rates is greater at 12 p.m. The rate of photosynthesis (100) is greater than the rate of respiration (70).

Step 2 Write the inequality for 12 p.m.: rate of photosynthesis > rate of respiration.

Step 3 Determine which of the rate is greater at 10 p.m. The rate of rate photosynthesis (0) is less than the rate of respiration (50).

Step 4 Write the inequality for 10 p.m.: rate of photosynthesis < rate of respiration.

2 **The below equation shows the relationship between distance from a light source and light intensity:**

$$\text{Light intensity} \propto \frac{1}{\text{distance}^2}$$

Write a sentence that summarises this relationship.

The symbol $\propto$ means 'proportional', so the relationship between distance from light source and light intensity can be summarised as:

Light intensity is proportional to one over the distance from the light source squared.

3 **During an enzyme investigation, 15.76 mg of product was produced. What whole milligram does this mass approximately equal?**

The mass 15.76 mg is approximately 16 mg, so:

$15.76\,\text{mg} \sim 16\,\text{mg}$

B Guided question

1 **Within a certain range of temperatures, the rate of decomposition in soil is proportional to the soil temperature. Write an expression to show this relationship.**

Step 1 Write down the two factors with a space between them and consider which of the symbols best fits.

Rate of decomposition .... soil temperature

C Practice questions

2 Write an inequality that compares the blood pressure in arteries and veins.

3 Write an expression that links rate of reaction and enzyme concentration when substrate concentration is not a limiting factor.

Rearranging the subject of an equation

★ **For WJEC/Eduqas GCSE Double Award, this is only required for HT students.**

A lot of the equations in GCSE Science, and Physics especially, only have three variables.

For most three-variable equations, it is possible to use the 'magic triangle' to change the subject of an equation.

For example:

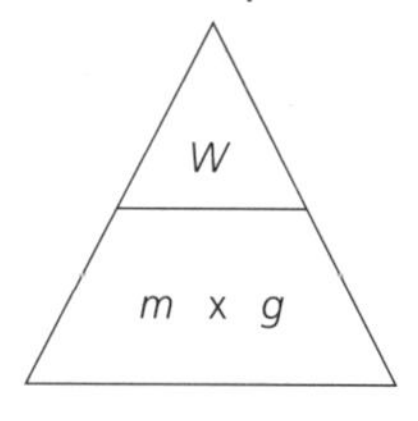

▲ Figure 1.8

- To make W the subject, cover up the W with your thumb to reveal $W = m \times g$
- To make m the subject, cover up the m with your thumb to reveal $m = \frac{W}{g}$
- To make g the subject, cover up the g with your thumb to reveal $g = \frac{W}{m}$

Magic triangles can be useful, but it is much better to develop the mathematical skills you need to solve equations without them. That way you can apply these transferable skills to solve a much wider range of problems.

If you have more than three variables, or you do not want to use a magic triangle, keep things simple. Remember:

- take one step at a time
- what you do to one side of an equation, you must also do to the other.

In other words, if you multiply one side of an equation by a value (whether the value is a number or a letter) you need to do the same to the other side.

For example, in the hypothetical equations below you could rearrange them as follows:

① $x = y + z$

If you wanted the subject to be z instead of x, you need to subtract y from both sides:

$$x - y = + z \equiv z = x - y$$

② $x = y \times z$

Tip

Remember that the skills you need to use algebra and solve equations in science are no different to those you use in maths. Don't let the different context distract you from applying the skills you already have.

If you wanted the subject to be y instead of x, you need to divide by z on both sides.

Remember that dividing by z on the right-hand side will cancel out the multiplying by z:

$$x \div z = y \equiv y = x \div z$$

Similarly, if you wanted the subject to be z in this equation, you would divide by y on both sides.

A Worked examples

1 **One of the equations of motion is $v^2 = u^2 + 2as$.**

Rearrange this equation to make s the subject.

Step 1 Work out what the question is asking: at the moment, v^2 is the subject, because the equation starts $v^2 = \ldots$ We have to change this to be $s = \ldots$

Step 2 Subtract u^2 from both sides: $v^2 - u^2 = u^2 + 2as - u^2$

Step 3 Simplify the right-hand side (RHS): $v^2 - u^2 = 2as$

Step 4 Divide both sides by $2a$: $\frac{v^2 - u^2}{2a} = \frac{2as}{2a}$

Step 5 Simplify the RHS: $\frac{v^2 - u^2}{2a} = s$

Step 6 Switch RHS and LHS: $s = \frac{v^2 - u^2}{2a}$

2 **Make mass the subject of this equation, moles $= \frac{\text{mass}}{M_r}$**

Step 1 Switch sides to get the new subject on the left.

$\frac{\text{mass}}{M_r} = \text{moles}$

Step 2 To get mass by itself on the left-hand side, you need to remove M_r by multiplying both sides by M_r and cancelling the M_r on the left.

$\frac{\text{mass} \times \cancel{M_r}}{\cancel{M_r}} = \text{moles} \times M_r$

So, the answer is mass=moles$\times M_r$

B Guided question

1 **The energy, E, stored in a stretched spring is given by the equation $E = \frac{1}{2}kx^2$**

Rearrange the equation to make x the subject.

Step 1 Multiply both sides by 2: $2E =$ ....

Step 2 Divide both sides by k: $\frac{2E}{k} = \frac{\ldots\ldots\ldots}{k}$

Step 3 Simplify: $\frac{2E}{k} =$ ....

Step 4 Take the square root of both sides: $\sqrt{\frac{2E}{k}} =$ ....

Step 5 Switch RHS and LHS: $x =$ ....

Tip

For squaring (^{2}) and square rooting ($\sqrt{\ }$) you also need to perform the same action to both sides, just as you would for any other mathematical function (adding, subtracting, dividing or multiplying).

2 Make volume the subject of the equation, moles = $\frac{(\text{volume} \times \text{conc.})}{1000}$

Step 1 Switch sides to get the new subject on the left

$$\frac{(\text{volume} \times \text{conc.})}{1000} = \text{moles}$$

Step 2 To get volume by itself as the subject on the left-hand side, you need to multiply both sides by 1000 and simplify

$$\frac{(\text{volume} \times \text{conc.} \times \cancel{1000})}{\cancel{1000}} = \text{moles} \times 1000$$

Step 3 To get volume by itself you now need to divide both sides by the conc.

....

C Practice questions

3 Rearrange the following equations to make x the subject.

a $y = 2x + 1$ b $3x = 4 + y$ c $y = mx + c$ d $2y + 3 = 4 - x$

4 Rearrange the equations below to make the variable in bold the subject.

a percentage yield = $\frac{\text{actual yield} \times 100}{\textbf{theoretical yield}}$

b volume = $\frac{\text{moles} \times \textbf{conc.}}{1000}$

c mean rate of reaction = $\frac{\text{quantity of reactant used}}{\textbf{time taken}}$

Substituting values into an equation

Once you have amended the subject of the equation to whatever you want to calculate (if necessary), the next step is to replace the letters with any values you have.

To solve an equation successfully, the key is to work carefully and logically through the steps, ensuring that all substitutions are done correctly and all the functions in the equation are evaluated accurately. It is important to double-check all your working as it is very easy to make mistakes in algebra.

You must use the correct units for measurement in an equation when substituting numerical values into it.

Tip

You may find it easier to substitute values into an equation *before* changing the subject. See what works best for you.

Tip

The most common mistake in substituting values is to substitute a value for the wrong letter. Take your time to check which letter stands for which value. If your final answer seems wrong or unrealistic, go back to see if you have made this mistake.

A Worked examples

1 The power of an electric drill is 250 W. If the drill is switched on for 40 seconds, how much work is done?

Step 1 State the equation (with the correct subject): $W = P \times t$

Step 2 Substitute in the values for power and time: $W = 250 \times 40$

Step 3 Carry out the calculation: $W = 10\,000$ J

2 Calculate the amount, in moles, of calcium hydroxide present in 25.0 cm^3 of a solution of concentration 0.25 mol/dm^3.

Step 1 The question gives a volume and a concentration and the number of moles is to be calculated. Hence, the equation to use is

$$\text{amount in moles} = \frac{\text{volume}\left(cm^3\right) \times \text{conc.}\left(mol/dm^3\right)}{1000}$$

Step 2 The subject is amount in moles, so does not need to be changed. Simply substitute the numerical values and calculate the answer.

$$\text{amount in moles} = \frac{25.0 \times 0.25}{1000} = 0.0063 \text{ to 2 significant figures}$$

Tip

If the volume of calcium hydroxide was given in dm^3 then you use the equation moles = volume × conc. and do not divide by 1000.

B Guided question

1 Calculate the concentration of calcium hydroxide solution obtained when 0.0034 mole of calcium hydroxide is dissolved in 15.0 cm^3 water.

Step 1 The information in the question is moles = 0.0034 mol and volume = 15.0 cm^3. Hence, use the equation:

$$\text{amount in moles} = \frac{\text{volume}\left(cm^3\right) \times \text{conc.}\left(mol/dm^3\right)}{1000}$$

Step 2 The subject is amount in moles, but you need to find the concentration, so the subject needs to be changed to conc.

Switch the sides so that conc. is on the left.

$$\frac{\text{volume}\left(cm^3\right) \times \text{conc.}\left(mol/dm^3\right)}{1000} = \text{amount in moles}$$

Step 3 You require conc. on its own on the left. To remove the 1000 multiply both sides by 1000 and simplify.

$$\frac{\text{volume} \times \text{conc.} \times \cancel{1000}}{\cancel{1000}} = \text{moles} \times 1000$$

$$\text{volume} \times \text{conc.} = \text{moles} \times 1000$$

Step 4 To get conc. on its own on the left divide both sides by volume and simplify.

Step 5 Then substitute the numerical value: moles = 0.0034 mol and volume = 15.0 cm^3, and calculate the answer.

2 A car accelerates from rest to 30 m/s in 12 s. Find its acceleration.

Step 1 State the equation: $a =$ ....

Step 2 Substitute in the values: $a =$ ....

Step 3 Carry out the calculation: $a =$ .... m/s^2

C Practice question

3 An investigation was carried out into effectiveness of different antibiotics, by comparing the areas of clear zones produced on bacterial cultures. The clear zones were approximately circular. The equation to calculate the area of the clear zones is shown below:

$$\text{Area of clear zone} = \pi r^2$$
$$r = \text{radius}$$

The radius of one of the clear zones was 17 mm. Use the formula to calculate the area of this clear zone.

Solving simple equations

Solving an equation means carrying out the final calculation. If you use a calculator, it is unlikely you will get this calculation wrong. However, you might still make a mistake rearranging or substituting, so make sure you take care with all previous steps before solving.

Remember that, in mathematical questions, examiners generally look for a formula, correct substitutions, a correctly calculated answer and – if required – a unit. So it is a good idea to substitute values for letters as soon as possible. If you try to rearrange the letters in the equation and get it wrong, it could cost you the substitution, arithmetic and final answer marks.

C Practice questions

1 In a chromatography experiment the distance moved by the solvent is 10.2 cm. Calculate the distance moved by a substance if its R_f value is 0.80.

2 Calculate the percentage atom economy for making copper(II) sulfate from copper carbonate and sulfuric acid.

$$CuCO_3 + H_2SO_4 \rightarrow CuSO_4 + H_2O + CO_2$$

Relative formula masses: $CuCO_3 = 124$, $H_2SO_4 = 98$, $CuSO_4 = 160$, $H_2O = 18$, $CO_2 = 44$

3 The speed of orange light in air is 3×10^8 m/s and its frequency is 5×10^{14} Hz. Find its wavelength, giving your answer in standard index form.

4 In a food chain, the energy available to the primary consumers can be calculated using the equation:

Energy available to primary consumers = energy in primary producers – energy lost in respiration – energy lost by waste and death

The energy available to primary consumers is 20 000 kJ, the energy lost in respiration is 30 000 kJ and the energy lost by waste and death is 150 000 kJ. What was the energy in the primary producers?

Key terms

Direct proportion: Quantities x and y are directly proportional to each other if their ratio $y : x$ is constant.

Inverse proportion: Quantities x and y are inversely proportional to each other if their product xy is constant.

Inverse proportion

Direct proportion and inverse proportion can also appear in algebraic equations. Remember that inverse proportion occurs when doubling one quantity causes the other quantity to halve.

★ **Only explicitly required for CCEA GCSE Science.**

For example, the equation for pressure is $P = \frac{F}{A}$. If the force F remains constant, then doubling the area A will cause the pressure P to halve.

Here are some values of pressure and area:

Pressure (N/m²)	120	60	40	30	20
Area (m²)	1	2	3	4	6

You can see that the pressure is decreasing while the area is increasing. This is the first hint that there is inverse proportion. The second hint is that, when we double the area, the pressure halves. However, the conclusive test is to check the product (which is what we get when we multiply pressure and area together). In this case, the product is always 120, so we can say that the pressure is inversely proportional to the area.

A Worked example

The resistance of five wires is measured. All are made of the same material and have the same length, but they have different cross-sectional areas. The results are shown in the table.

Resistance R (Ω)	60	30	20	15	10
Area *A* (mm²)	0.5	1.0	1.5	2.0	3.0

a **Show that the resistance is inversely proportional to the cross-sectional area of the wire.**

Step 1 Find the product, *RA*, for each wire:

Resistance R (Ω)	60	30	20	15	10
Area *A* (mm²)	0.5	1.0	1.5	2.0	3.0
RA (Ωmm²)	30	30	30	30	30

Step 2 Since the product *RA* is constant, we can confirm that the resistance is inversely proportional to the cross-sectional area of the wire.

b **Calculate the resistance when the area is 2.5 mm².**

Step 1 Use the information we know about the product, *RA*: $R \times A = 30$

Step 2 Substitute for *A*: $R \times 2.5 = 30$

Step 3 Divide both sides by 2.5 and solve: $R = \frac{30}{2.5} = 12\,\Omega$

Note, it may be tempting to look at the table and think that, since 2.5 is half way between 2.0 and 3.0, the resistance should be half way between 15 and 10 Ω (12.5 Ω). This would be incorrect.

B Guided question

1 **The power *P* of the electric element in a domestic iron is inversely proportional to its resistance *R* when voltage is constant. When the resistance is 48 Ω, the power is 1200 W.**

Calculate the power if the element is replaced with a 60 Ω element.

Step 1 Since power *P* is inversely proportional to *R*, then $PR = a$ ....

Step 2 In this case, with the 48 Ω element, $PR =$ .... × .... = ...

Step 3 With the 60 Ω element, $57\,600 = P \times$ ....

Step 4 Solve: $P = \frac{57\,600}{\text{..........}} =$ .... W

C Practice question

2 The intensity I of the light received from a lighthouse is inversely proportional to the square of its distance from the observer d^2. An experiment for a particular lighthouse gives the following data:

Intensity (W/m^2)	720	360	240
d^2 (m^2)	2	4	6

The intensity of the light received by a ship is 0.001 W/m^2.

a Use the table to find the value of d^2 for this ship.

b Use your answer to part a to show that the distance between the ship and the lighthouse is 1200 m.

» Graphs

Science students are often asked to plot or interpret graphs arising from experimental data. This can take a number of different forms, including reading values off a graph or finding gradients and intercepts.

After carrying out an experiment, it is often useful to plot a graph to help analyse your results. A graph is an illustration of how two variables relate to one another.

> **Tip**
> Use a ruler to read points off a graph. This helps ensure that you do not make mistakes.

General construction

When drawing graphs, always:

- use graph paper for accuracy (otherwise your graph is just a sketch)
- draw it in pencil and rule the axes with a ruler (and have an eraser handy)
- label your axes, giving units where appropriate
- choose a suitable scale to make sure you use as much of the graph paper as possible – at the very least, half of the graph paper
- draw a line graph unless told otherwise
- draw a straight line or curve of best fit.

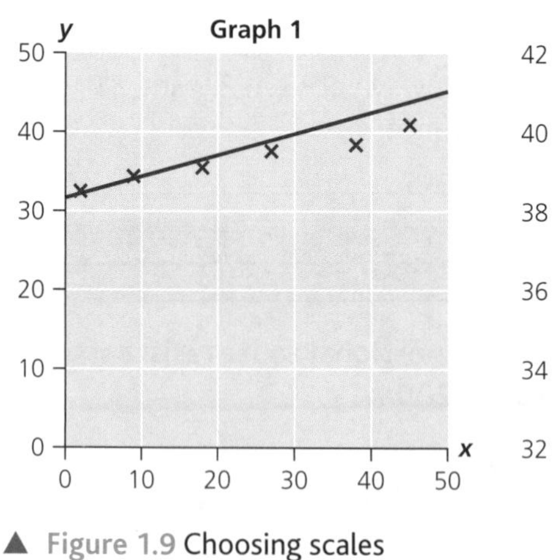

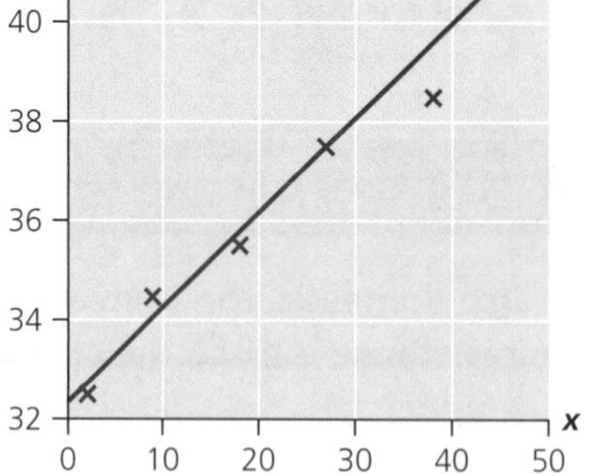

▲ Figure 1.9 Choosing scales

Graph 1 – a poor scale in the y direction that compresses the points into a small section of graph paper. ✗

Graph 2 – a good scale as the points fill more than half the graph paper in both x and y directions ✔

- It is also important when choosing a scale that you examine the data to establish whether it is necessary to start the scale(s) at zero. Graph 2 does not include the origin.
- Choose a simple scale increasing in multiples of 2, 5 or 10 – avoid using multiples of 3 or 7.

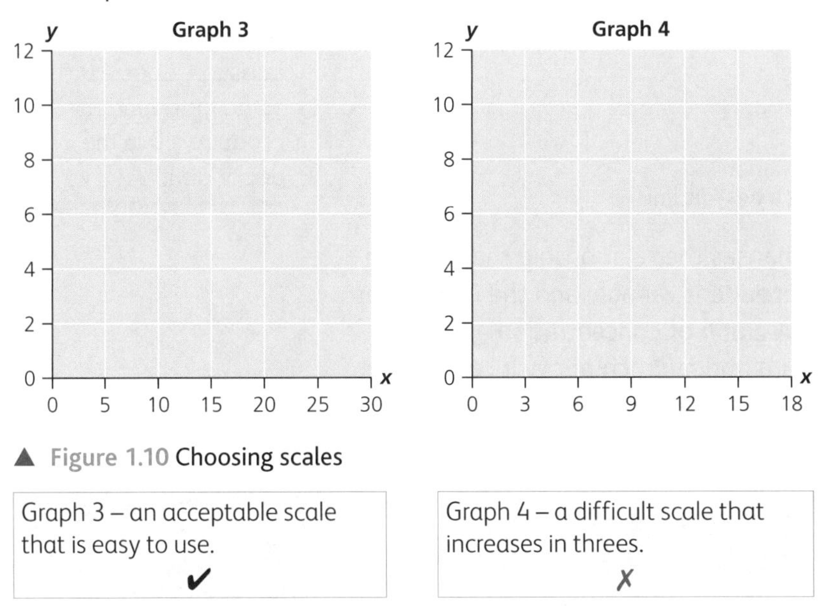

▲ Figure 1.10 Choosing scales

- The independent variable is placed on the *x*-axis and the dependent variable on the *y*-axis. Axes should be labelled with the name of the variables and the unit of measurement for each. For example, one of the labels may be temperature/°C or temperature in °C.
- Data points should be marked with a cross (✗) so that all points can be seen when a line of best fit is drawn.
- A line of best fit should be drawn. When judging the position of the line, there should be approximately the same number of data points on each side of the line; resist the temptation to simply connect the first and last points. The line of best fit can be either a straight line or a curve.

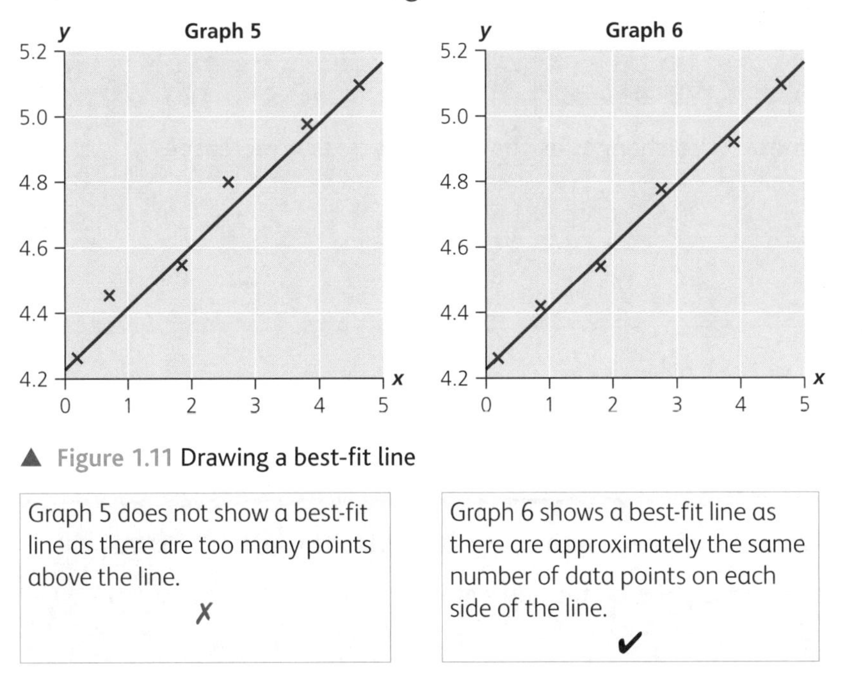

▲ Figure 1.11 Drawing a best-fit line

Tip

A line of best fit is added by eye. You should use a transparent plastic ruler or a flexible curve to aid you.

- When drawing a best-fit line or curve, ignore any anomalous results.

▲ Figure 1.12 Ignore anomalies when drawing a best-fit line

> **Tip**
> Not all lines of best fit go through the origin – before using the origin as a point always ask the question 'does a 0 in the independent produce a 0 in the dependent?

- The graph should have a title that summarises the relationship that is being illustrated – this should include the independent variable and the dependent variable. For example, a suitable title is 'A graph of concentration against time for the reaction between magnesium and hydrochloric acid' or simply 'A concentration–time graph for the reaction between magnesium and hydrochloric acid'.

Translating between graphical and numerical form

Most (but not all) of the graphs you will be asked to draw result in straight line graphs of positive gradient. There are two types of straight line graphs of positive gradient, each with a general equation.

- $y = mx$, where gradient is m and the line passes through the origin (0,0); this shows proportionality,
- $y = mx + c$, where gradient is m and the line passes through the y-axis at a point (0,c); this shows a linear relationship but one that isn't proportional. See page 62 for more on this.

> **Tip**
> Students often think of graphs as having a straight line of best fit and a positive gradient, but remember that graphs can have a negative gradient, and that a line of best fit may be a curve.

You may also be asked to obtain numerical data from a graph in your exam. To do this, draw construction lines on the graph from the axes to meet the gradient as shown in the worked example. Use the lines to read off the value on the axes.

A Worked example

This graph shows how the current in a metal wire changes as the voltage across it increases.

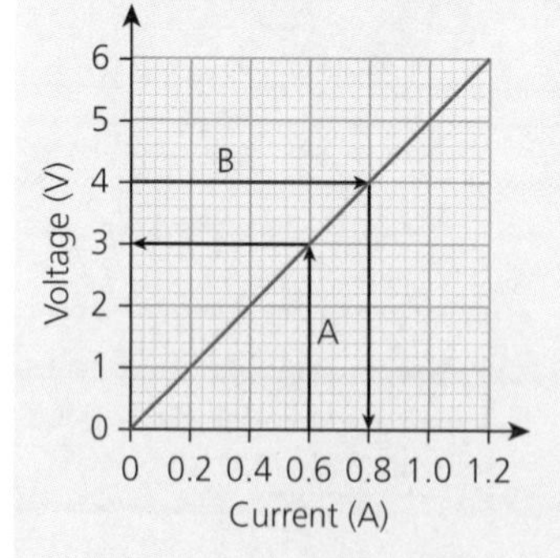

a Use the graph to find the voltage across the wire when the current flowing through it is 0.6A.

Step 1 Draw a vertical line (labelled A) from the point on the horizontal axis where the current is 0.6A, up to the gradient.

Step 2 Continue across from the gradient to where your horizontal line reaches the vertical axis. The reading on the vertical axis, 3V, is the answer.

b Use the graph to find the current in the wire when the voltage across it is 4V.

Step 1 Draw a horizontal line (labelled B) from the point on the vertical axis where the voltage is 4V, across to the gradient.

Step 2 Continue down from the gradient to where your vertical line reaches the horizontal axis. The reading on the horizontal axis, 0.8A, is the answer.

B Guided question

1 This graph shows how the speed of a cyclist changes with time.

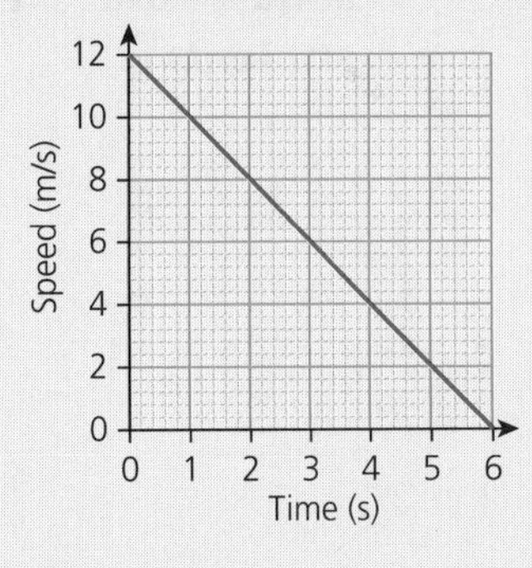

a At what time is the cyclist travelling at 7 m/s?

Step 1 This speed on the vertical axis is half way between .... m/s and .... m/s.

Step 2 At this speed, draw a .... line to the graph.

Step 3 From the point where this line meets the graph, draw a .... line to the time axis.

Step 4 The line meets the time axis at .... seconds. This is the answer.

b What is the speed of the cyclist when the time is 1.5 s?

Step 1 This time on the horizontal axis is half way between .... s and .... s.

Step 2 At this time, draw a .... line to the graph.

Step 3 From the point where this line meets the graph, draw a horizontal line to the .... axis.

Step 4 The line meets the speed axis at .... m/s. This is the answer.

C Practice question

2 In an experiment some calcium carbonate and acid were placed in a conical flask on a balance and the balance reading recorded every minute. The results were recorded, and the graph shown below was drawn.

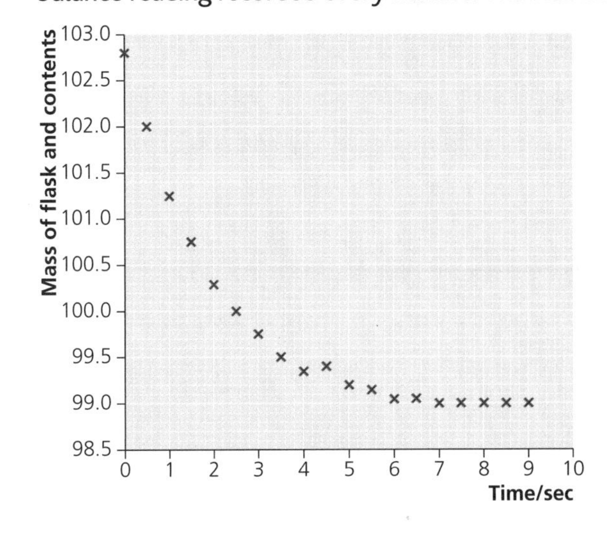

a Are there any results that you would ignore when drawing a best-fit curve?
b Look at the labels on the axes and describe any changes you would make.
c Suggest a title for this graph.
d Do you think the scale is appropriate in this graph? Explain your answer.

Understanding that $y = mx + c$ represents a linear relationship

★ **For WJEC/Eduqas GCSE Science Double Award, this is only required for HT students.**

As we have already seen, all straight line graphs can be written in the form $y = mx + c$. This shows a linear relationship where the graph of y against x is a straight line that does not go through the (0,0) origin.

In the equation $y = mx + c$:

- m = gradient of the line
- c = y intercept (the point where the line crosses the y-axis).

Therefore, if the line has a gradient of 2 and the y intercept is 0.1 the equation of the line would be: $y = 2x + 0.1$

This means that if $x = 4$ then y would be:

$$y = 2 \times 4 + 0.1$$

$$y = 8.1$$

Straight lines that do go through the (0,0) origin have the form $y = mx$. These lines are special because they show direct proportion.

Tip

A straight line through the origin is the quickest test for a relationship of direct proportion.

In your exams, you could be asked to sketch a graph of a linear relationship. As the graph is a straight line, only two points are needed to draw the line, although it is recommended that you use a third point to check whether the line is correct.

If a set of axes is given in the question, it does not particularly matter which values on the x-axis you use to construct the line, as long as their corresponding y values are within the range shown on the y-axis. However, using points fairly far apart could make the line easier to draw accurately.

If the question leaves it up to you to draw the axes, make sure that the scale you choose for each axis covers the range of values in the given data. Do this by finding the maximum and minimum y values before starting to draw the axes.

A Worked examples

1 Look at these graphs.

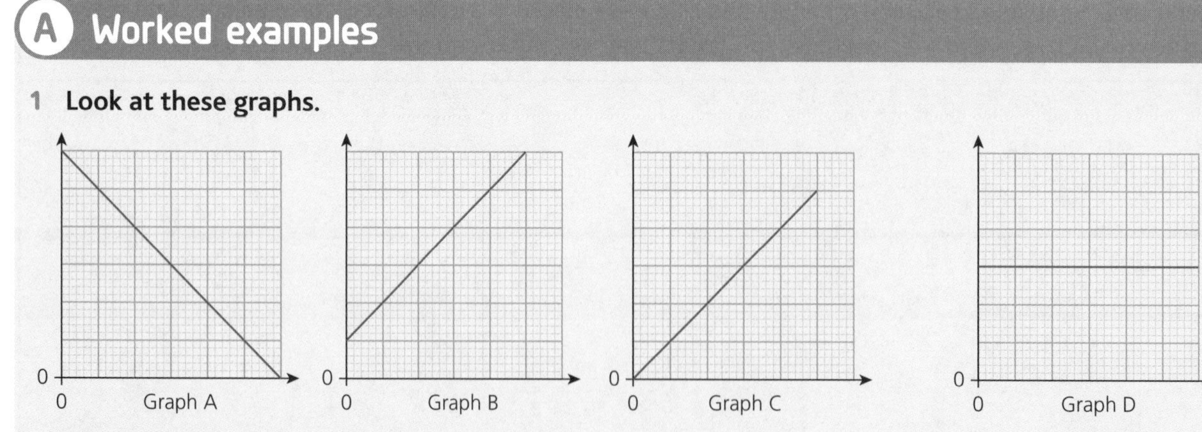

State which of the graphs show:

a a linear relationship **b direct proportion.**

Give reasons for your answers.

Step 1 All the graphs are straight lines – so they *all* show linear relationships.

Step 2 Only graph C is a straight line through the point (0,0), so only graph C shows direct proportion.

Step 3 Graphs A, B and D do not show direct proportion because they do not pass through (0,0).

2 **The effect of enzyme concentration on the rate of a reaction can be predicted by the equation $y = 2x + 4$. Sketch the graph of this relationship on the axes provided.**

Step 1 Identify the values of m and c in the linear equation. The given equation is of the form $y = mx + c$ with m = 2 and c = 4.

Step 2 Choose two x values (within the range shown on the given x-axis) and calculate their corresponding y values:

At $x = 1$, $y = 2 \times 1 + 4 = 6$

At $x = 5$, $y = 2 \times 5 + 4 = 14$

Step 3 Plot these points on the set of axes and draw a straight line through them.

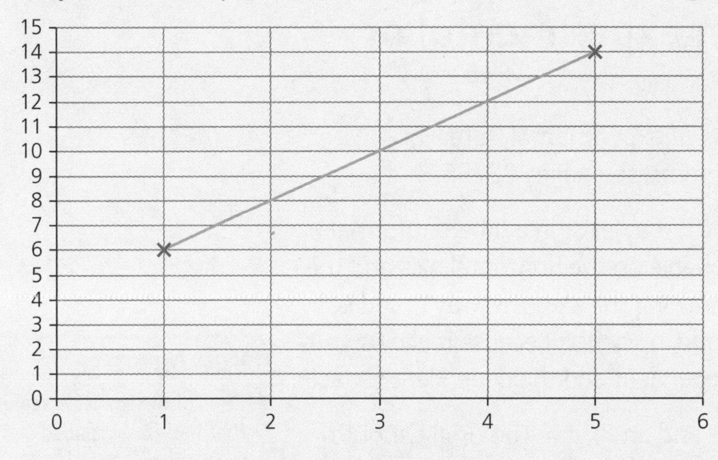

B Guided question

1 **Sketch the graph of $y = -0.5x + 9$ on the axes provided.**

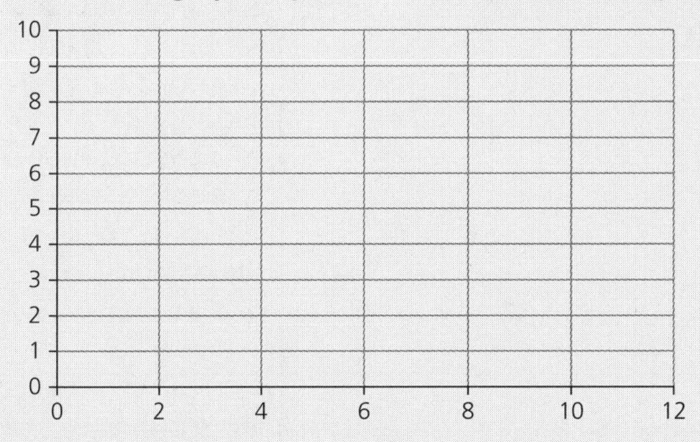

Step 1 Identify the values of m and c in the linear equation.

In this equation, m = ... and c = ...

Step 2 Choose two x values (within the range shown on the given x-axis) and calculate their corresponding y values.

At $x = 0$, $y =$ ....

At $x = 10$, $y =$ ....

Step 3 Plot these points on the set of axes and draw a straight line through them.

Tip

As the gradient (m) in this question is negative, the graph is a straight line with negative gradient, so it will slope downwards from left to right.

C Practice question

2 Sketch the graph of $y = 3x + 4$

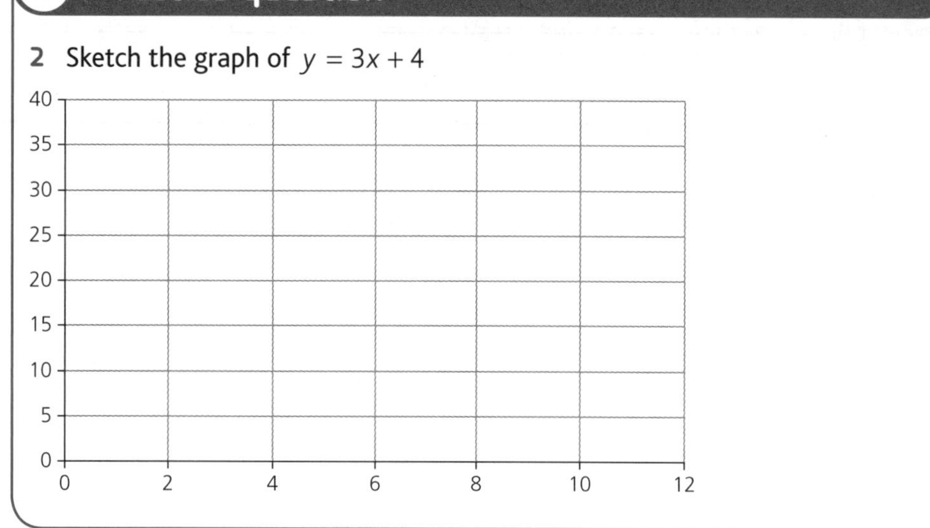

Plotting two variables from experimental or other data

This section will cover how to plot line graphs of experimental data. For information on plotting bar charts and histograms, see pages 32–36.

When plotting line graphs, take your data – which may exist in a different format (such as a table) – and plot the independent variable on the horizontal axis and the dependent variable on the vertical axis. While drawing the axes this way round is not essential, it is the way that it is normally done in science, because it allows you to see clearly the relationship between the independent and the dependent variable.

Ensure that each axis has a continuous scale and an origin. The origin should be appropriate for a graph, but does not have to be zero or be the same for both axes. For example, if a data set used to draw a graph runs from 100–200 then it would be logical not to have zero as an origin; 90 or 100 could be used instead.

Key terms

Continuous scale: A scale that has equal spaced increments.

Origin: The start of an axis of a graph.

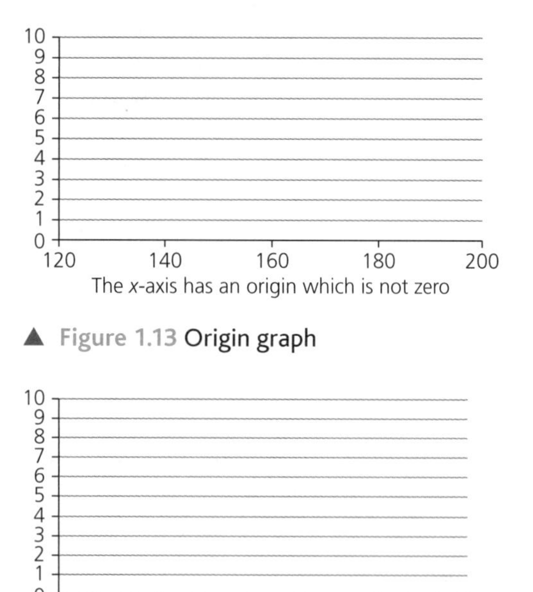

▲ Figure 1.13 Origin graph

0
1
2
3
4
5
6
7
8
9
10
1980
1984
1988
1990
1992
4 years
2 years

The x-axis is not a continuous scale because it does not go up in equal increments. This would be incorrect.

▲ Figure 1.14 Non-continuous scale graph

Tip

A common mistake in answering these types of questions is to have a scale on the horizontal axis that is not linear (for example does not increase by the same value from each marked number to the next).

A Worked examples

1 **In an experiment a lump of calcium carbonate was added to 50 cm^3 of hydrochloric acid in a conical flask and placed on a balance. A stopwatch was started as soon as the calcium carbonate made contact with the acid, and the mass was recorded every 20 seconds in the table below. Plot a graph of mass (*y*-axis) against time (*x*-axis).**

Time in s	0	20	40	60	80	100	120
Mass in g	234.10	233.70	233.40	233.20	233.05	233.00	233.00

Step 1 Decide on the scale for the *x*-axis. The graph paper has 12 squares across, hence it is appropriate to start at time zero and increase in intervals of 10 up to 120 seconds.

Step 2 Decide on the scale for the *y*-axis. The y-axis has 16 squares up. The masses range from 234.10 to 233.00, which is an interval of 1.1 g. It is not sensible to start this scale at zero as this would not spread out the data points over the graph paper. Instead begin at 232.80 and each square could represent 0.10 g. Count the number of zeros after the 1.

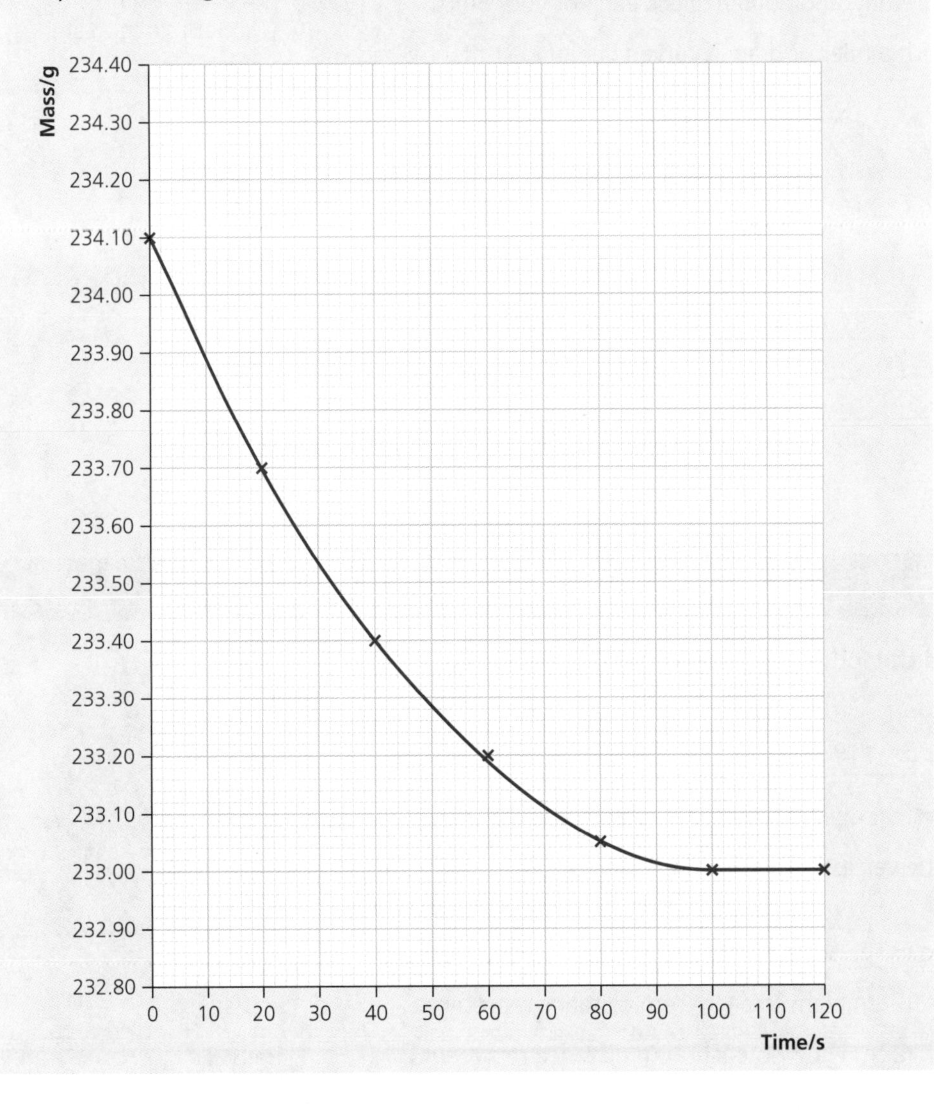

2 **The table below shows the results of a study into the rate of transpiration over a period of 24 hours.**

Time (hours)	Rate of transpiration (arbitrary units)
0	1
4	4
8	10
12	15
16	8
20	3
24	2

Plot the data on a graph.

Step 1 Draw suitable axes. These should be continuous, and have an origin. A zero origin for both axes can be used in this case.

Step 2 Label the axes with the correct headings. Use the headings from the table as your axes titles.

Step 3 Plot the points carefully, and double check each of your plots.

Step 4 Join the points with a ruler or draw a curved line of best fit.

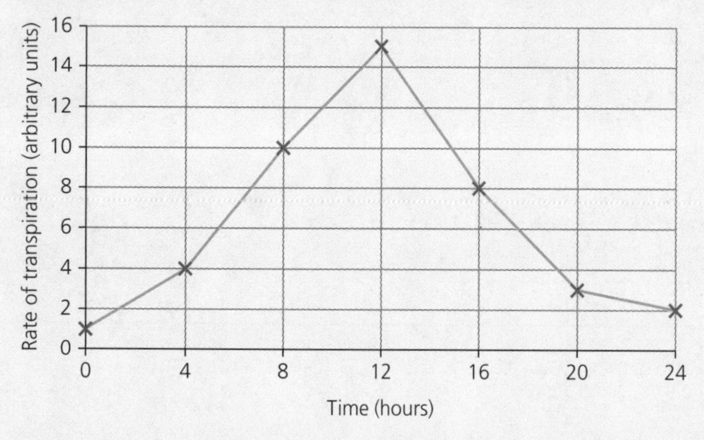

Tip

Some exam boards prefer you to draw curved lines of best fit, whereas some prefer the points to be joined with straight lines. Check with your teacher what you should do in exam questions.

B Guided question

1 **An experiment obtains the following data for variables x and y.**

x	0	1	2	3	4	5
y	4.5	6.0	7.5	9.0	10.5	12.0

Plot the graph of y (vertical axis) against x (horizontal axis) using this data.

Step 1 Draw and label the vertical axis with the letter .... and horizontal axis with the letter ....

Step 2 Decide on the scale. The scale must be linear and cover at least half the grid.

For the y-axis, the grid is 12 cm high, so each 1 cm distance represents .... unit(s).

For the x-axis, the grid is 12 cm wide, so each 1 cm distance represents .... unit(s).

Step 3 The first point is at the intersection where the vertical line at $x = 0$ meets the horizontal line at $y = 4.5$. The second point is at the intersection where the vertical line at $x = 1$ meets the horizontal line at $y =$ ....

Step 4 Repeat until all points are plotted.

C Practice questions

2 In an experiment, magnesium was reacted with sulfuric acid and the volume of hydrogen produced collected and measured, every 10 seconds, in a gas syringe.

Use the data in the table to draw a graph with best-fit curve of volume of hydrogen against time.

Time /s	0	10	20	30	40	50	60	70	80	90	100
Volume of hydrogen /cm^3	0	30	55	75	88	98	102	104	104	104	104

3 During an equilibrium reaction a gas C is produced from the reaction of gases A and B.

$$A(g) + B(g) \rightleftharpoons C(g)$$

The percentage of C in the reaction mixture varies with temperature. Plot a graph of percentage of C against temperature.

Temperature /°C	100	200	300	400	500
Percentage of C in equilibrium mixture / %	58	42	30	21	16

4 The following table shows the results of a survey of MRSA cases in an area over a period of time.

Time (months)	Number of MRSA cases
0	50
1	200
2	140
3	195
4	130

Plot the data on a graph.

Determining the slope and intercept of a straight line

The equation of a graph showing a linear relationship is $y = mx + c$ and the equation of a graph showing direct proportion is $y = mx$. In these equations, m represents the slope (gradient) of the line, and c represents the intercept on the y-axis. In your exam you may need to identify both the gradient and the intercept.

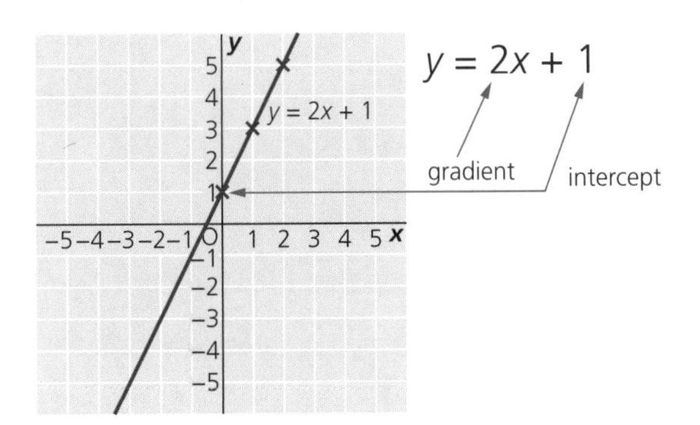

▲ Figure 1.15 Straight-line graph

Key terms

Gradient: This is another word for 'slope'. It is the change in the y-value divided by the change in the x-value.

Intercept: This is the point where the graph crosses an axis. In the equation: $y = mx + c$, the y-intercept is where the graph crosses the y-axis when $x = 0$; in other words, it is the value for y when $x = 0$.

Gradient is another word for 'slope'. The higher the gradient of a graph at a point, the steeper the line is at that point. A positive gradient means the line slopes up from left to right. A negative gradient means that the line slopes downwards from left to right. For a straight-line graph, the gradient is a constant value. A zero gradient graph is a horizontal line.

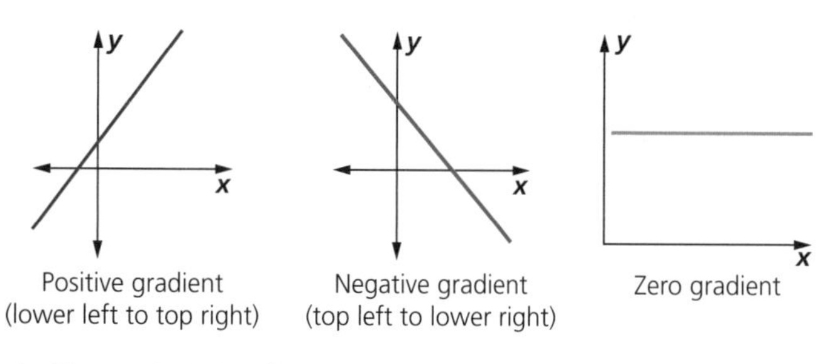

▲ Figure 1.16 Gradients

Determining intercept

Determining the intercept is usually fairly straightforward. In exam questions, you would normally be asked to find the intercept on the x-axis. To do this, either read off the x value at which the line or curve crosses the x-axis, or if the crossing point is not shown on the graph, you may be able to extrapolate from the line to find where it intersects the axis.

Key terms

Extrapolate: Extending a graph to estimate values.

Hypotenuse: The longest side of a right-angled triangle.

Determining slope/gradient

When provided with a graph showing a linear relationship, you can calculate the rate of change by finding the slope (gradient) of the line. This is an important skill in science because it allows the calculation of a variety of different rates, including rates of reaction.

To find the gradient of a line, divide the change in the variable on the y-axis by the corresponding change in the variable on the x-axis. It is easiest to determine the amount of change in each variable by drawing a right-angled triangle with its hypotenuse along the line. Because the gradient is the same at all points on a line, it does not matter where on the line you place this triangle.

Tip

It may be easier to remember gradient as

$$\text{slope} = \frac{\text{rise (height of m)}}{\text{run (length of m)}}$$

You can also write this equation as: gradient (m) $= \frac{\text{change in } y\text{-axis}}{\text{change in } x\text{-axis}} = \frac{\Delta y}{\Delta x}$

In the following example, this is: $\frac{100}{10} = 10$.

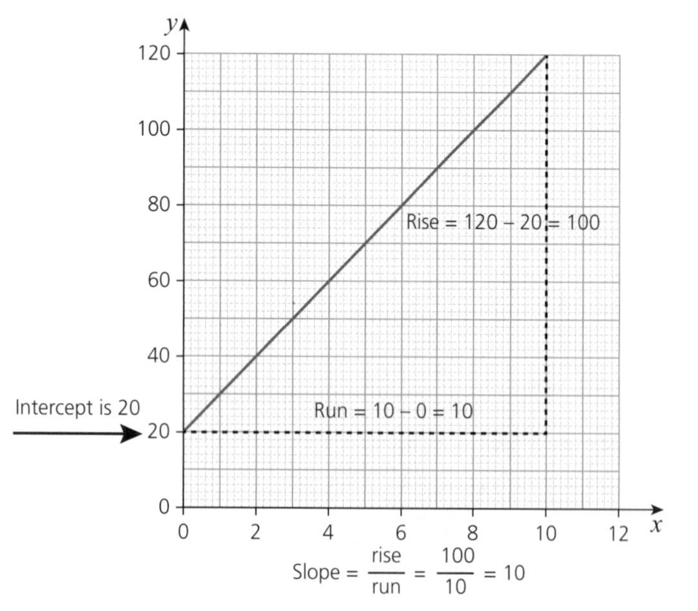

▲ Figure 1.17

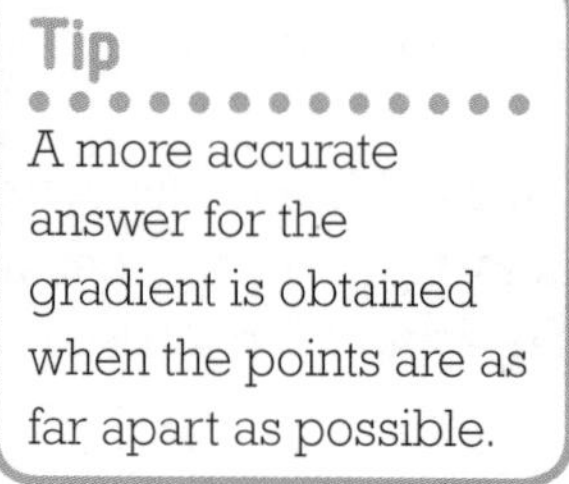

A Worked examples

1 **The graph below shows the volume of methane in a biogas generator over time. What is the rate of change of methane volume? Give your answer in m³/hour.**

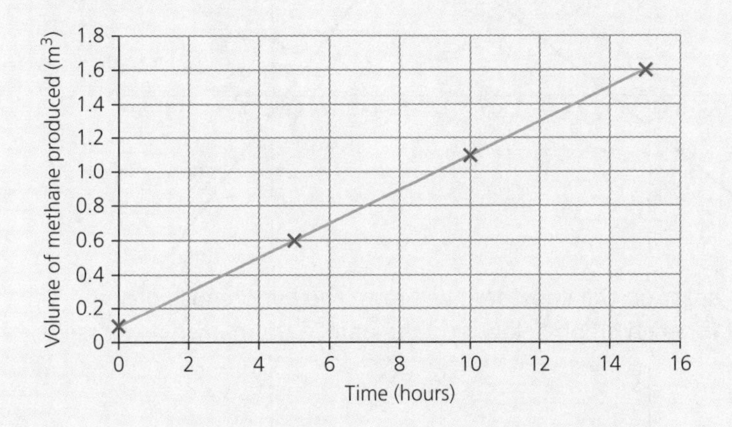

The rate of change is the gradient of the line.

Step 1 To find the gradient, draw a right-angled triangle as shown below. The triangle has a vertical edge and a horizontal edge, and its hypotenuse (slanted edge) lies along the line graph.

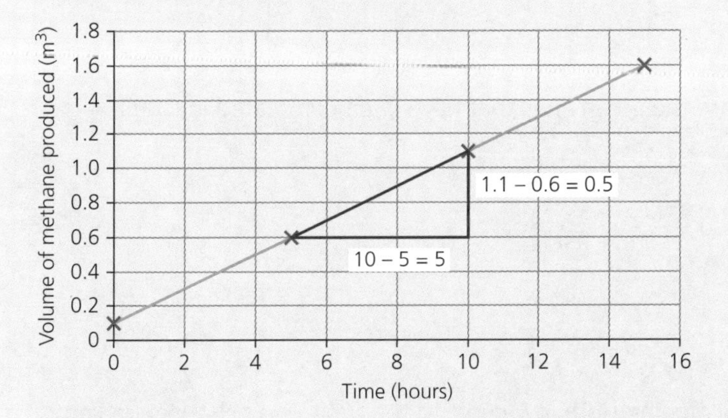

Step 2 Use the triangle to determine the change in x and the change in y:

Change in y = length of vertical edge of triangle

Change in x = length of horizontal edge of triangle

Step 3 Substitute these values into the equation below:

Gradient = change in y ÷ change in x

$= (1.1 - 0.6) \div (10 - 5) = 0.5 \div 5$

$= 0.1\,\text{m}^3/\text{hour}$

Tip

You should always show your working out when calculating the gradient of a line. It is also a good idea to show and label the triangle used.

B Guided question

1 a **Find the gradient of the lines in A, B and C.**

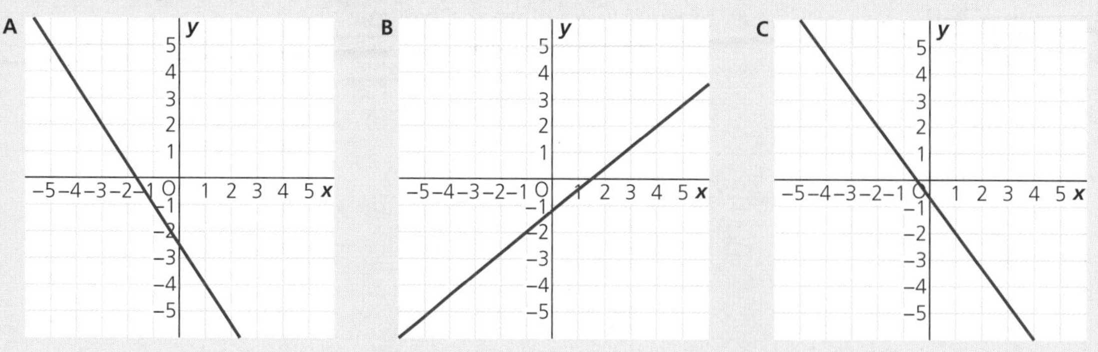

Step 1 Choose two points that are far apart on the line; these will form the hypotenuse of the triangle. This step has been completed for each graph below, and the points are marked as crosses.

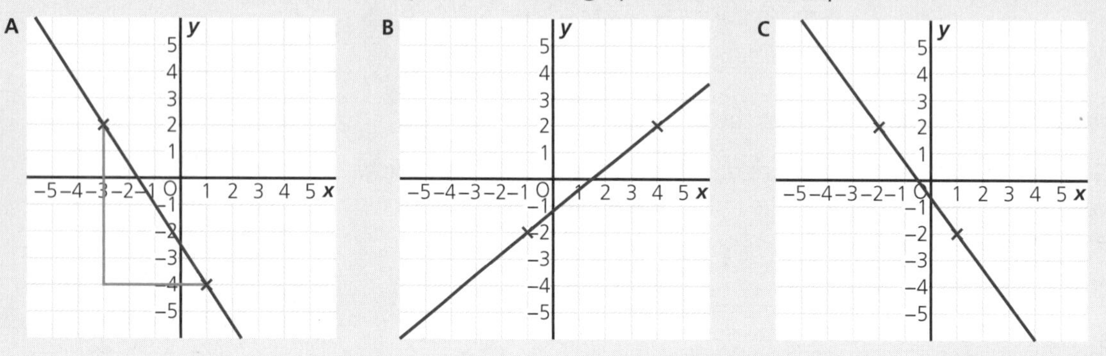

Step 2 Complete the triangle – these are the green lines that have been completed in Graph A

Step 3 Find the Δy (rise) value

Step 4 Find the Δx (run) value

Step 5 Find the gradient using the equation

$$\text{gradient (m)} = \frac{\text{change in } y\text{-axis}}{\text{change in } x\text{-axis}} = \frac{\Delta y}{\Delta x}$$

Step 6 Decide if it is a positive gradient sloping from lower left to top right or a negative gradient.

b **Find the y-intercept and write the equation for the line.**

Step 1 Write down the number at which the blue line cuts through the y-axis (at $x=0$). This is the intercept c.

Step 2 Substitute the values for m and for c into the equation $y = mx + c$.

C Practice questions

2 The graph below shows the results of an investigation into osmosis in a sample of onion tissue. What was the internal concentration of the onion cells?

3 The graph below shows the results of an investigation into the breakdown of starch by amylase. What is the fastest rate of reaction achieved during this investigation? Explain how you arrived at your answer.

Determining area under a straight-line graph

In some graph questions, you may also be asked to calculate the area below the line. This is the area between the graph and the *x*-axis. In this example you need to calculate the area of the triangle and add the area of the rectangle below the dotted line created by drawing the triangle.

To calculate the area of the triangle: $\frac{1}{2} \times \text{height} \times \text{length}$

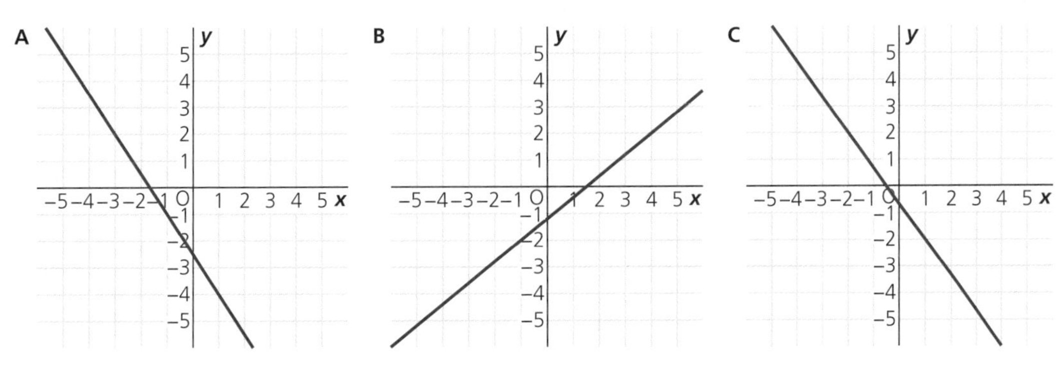

▲ Figure 1.18

In this example it is: $\frac{1}{2} \times 100 \times 10 = 500$.

To calculate the area of the rectangle below the dotted line: length × height

In this example it is: $10 \times 20 = 200$.

By adding these totals together you get: $500 + 200 = 700$.

We will look at how to find the slopes and area for curves next.

Drawing and using the slope of a tangent to a curve as a measure of rate of change

★ **Not explicitly required by CCEA GCSE Science Single or Double Award.**

When you plot data on a grid you must draw an appropriate line of best fit. At GCSE this will usually be a straight line. But, occasionally, you may have data that means your plots lie on a curve. You must draw a curve through as many points as you can.

The word tangent means 'touching' in Latin.

> **Key term**
>
> Tangent: This is a straight line that just touches the curve at a given point and does not cross the curve.

To draw a tangent at a point (x, y), follow these steps:

1. Place your ruler through the point (x, y).
2. Make sure your ruler goes through the point, and does not touch the curve at any other point.
3. Draw a ruled pencil line passing through point (x, y).

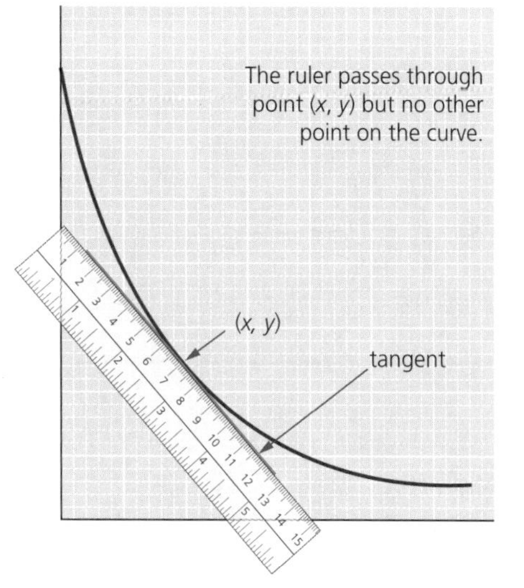

▲ Figure 1.19 Tangent to a curve

To calculate the gradient of a curve at a particular point it is necessary to draw a tangent to the curve at the point and calculate the gradient of the tangent.

A Worked example

In an experiment a student recorded the total volume of gas collected in a reaction at 20-second intervals.

Time /s	0	20	40	60	80	100	120
Volume of gas /cm³	0	21	42	56	65	72	72

a Plot a graph using the data shown and draw a line of best fit.

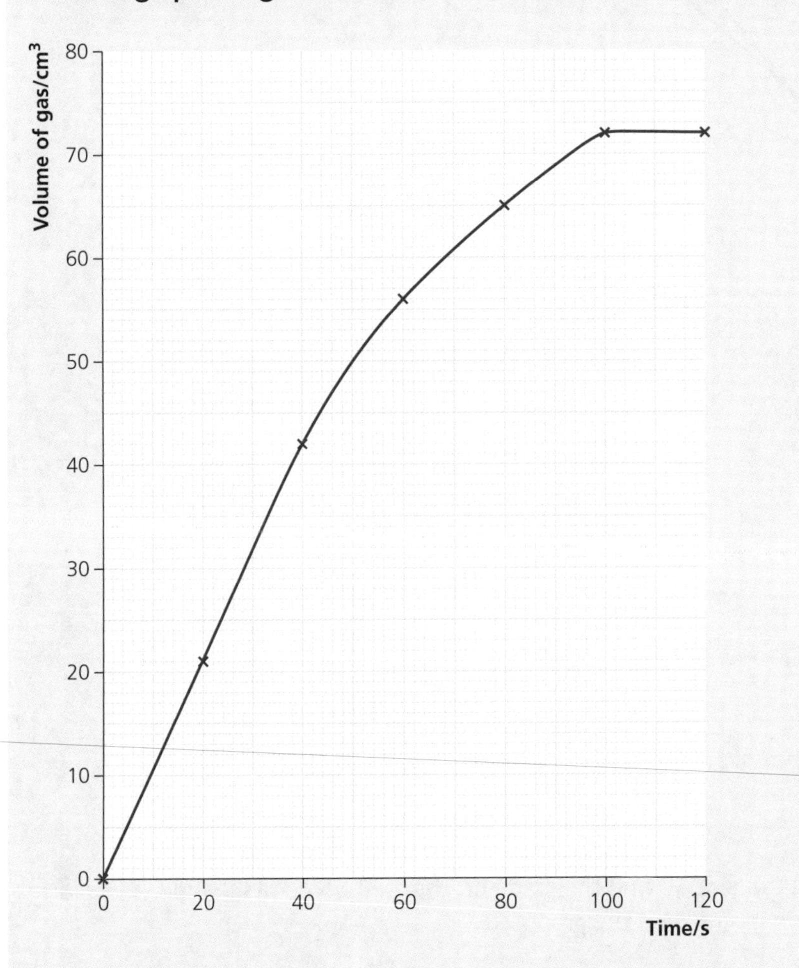

The graph will be a graph of volume of gas (y-axis) against time (x-axis). Allocating one large square for 20 s is a scale that is suitable on the x-axis, and one large square for 10 cm³ of gas is suitable on the y-axis.

b Use the graph to calculate the rate of reaction at 60 s in cm^3/s.

The graph is a curve and to find the rate of reaction at 60 s, a tangent to the curve must be drawn at 60 s as shown below in red.

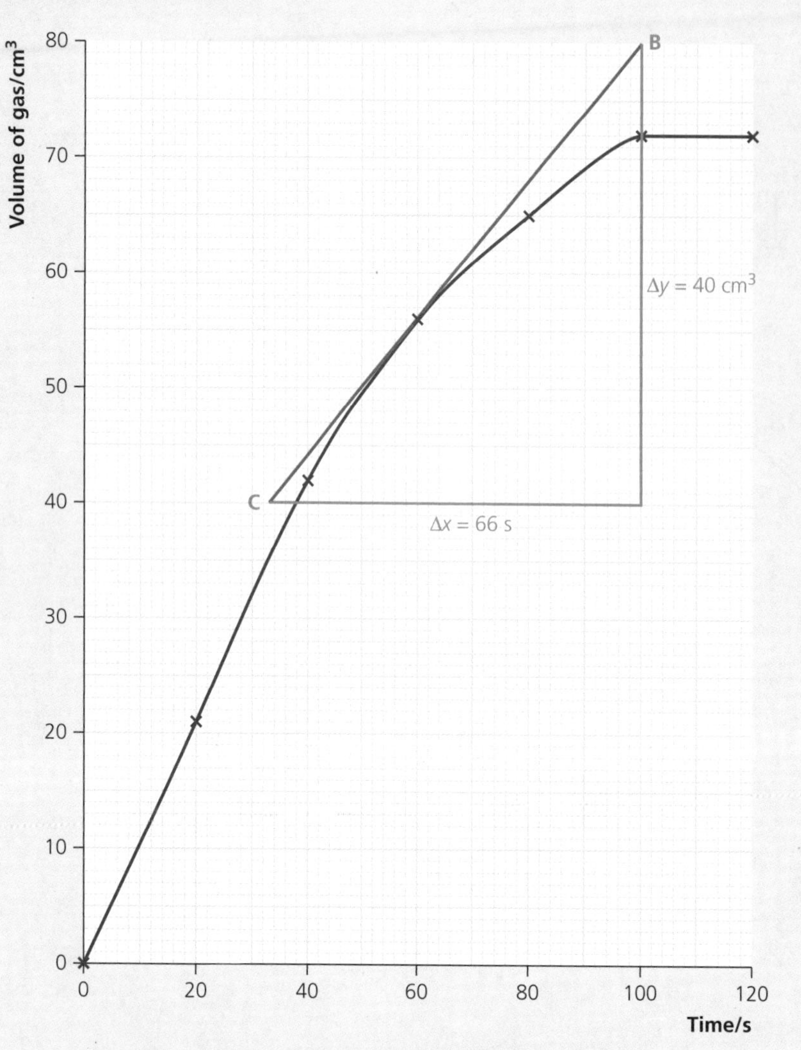

The rate is given by the gradient of this tangent. To find the gradient, choose two points B and C, far apart on the line, and form a triangle as shown in green on the graph.

$$\text{gradient (m)} = \frac{\text{change in } y\text{-axis}}{\text{change in } x\text{-axis}} = \frac{\Delta y}{\Delta x} = \frac{40}{66} = 0.61\,\text{cm}^3/\text{s (to 2 s.f.)}$$

B Guided question

1 **What is the gradient of the curve at point A?**

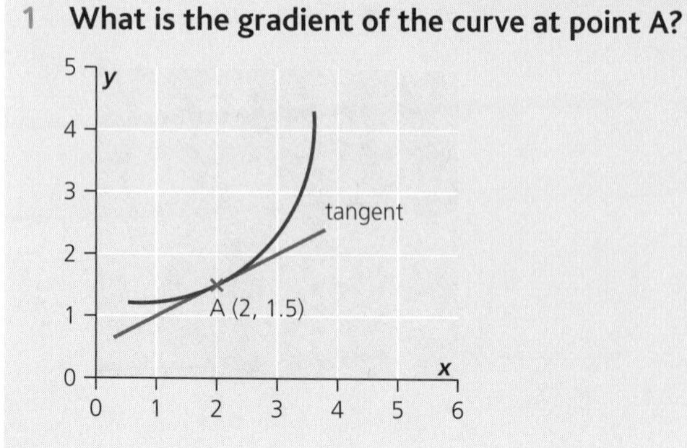

Step 1 To find the gradient of the curve at point A, a tangent to the curve at point A must be drawn. It is shown in red on the graph above.

Step 2 Choose two points, B and C, far apart on the line and form a triangle as shown in the graph below.

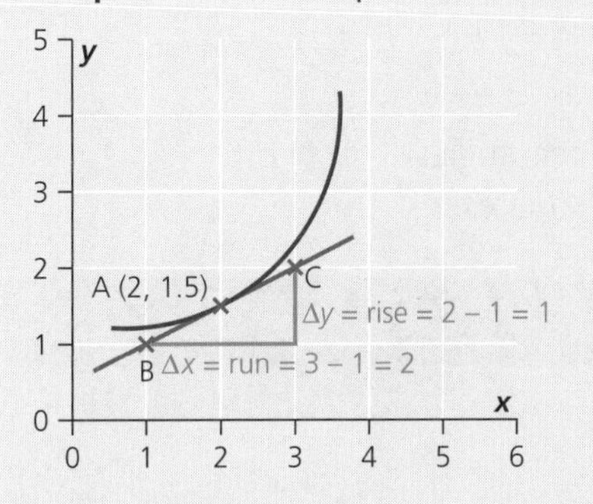

Step 3 Calculate the gradient using

$$\text{gradient (m)} = \frac{\text{change in } y\text{-axis}}{\text{change in } x\text{-axis}} = \frac{\Delta y}{\Delta x}$$

C Practice questions

2 Calculate the rate of reaction when the concentration of ester is 0.200 mol/dm³.

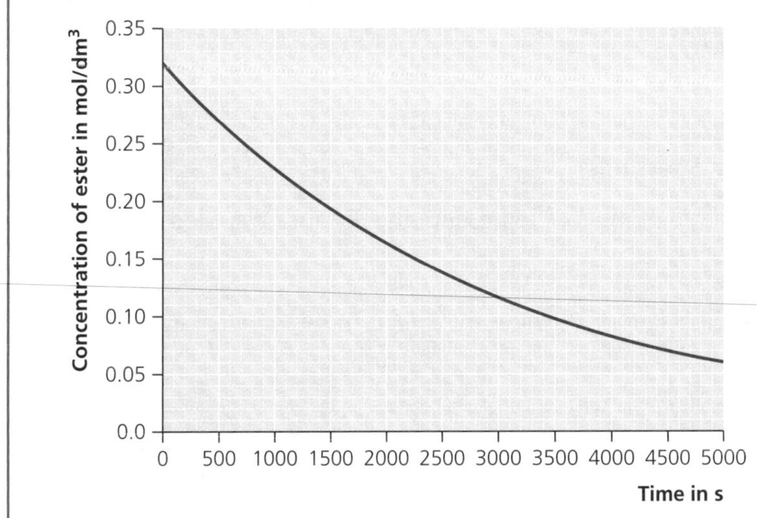

3 The volume of carbon dioxide gas produced over time when calcium carbonate reacted with hydrochloric acid was recorded in the table below.

Time / s	0	10	20	30	40	50	60	70	80	90	100
Volume of carbon dioxide / cm³	0	22	35	43	48	52	55	57	58	58	58

a Plot a graph of the volume of carbon dioxide against time.
b Calculate the rate of reaction at 20 seconds by drawing a tangent to the curve.
c Calculate the rate of reaction at 60 seconds by drawing a tangent to the curve.

Determining the area of a curve

We have seen that the gradient of a graph often has significance, but that in some graphs, the area between the graph and the horizontal axis is of greater interest.

★ **Not explicitly required by CCEA GCSE Science Single or Double Award.**

For example:

- the area between a speed-time graph and the horizontal axis represents the distance travelled
- the area between a force-extension graph and the horizontal axis represents the work done.

If the graph is a straight line, we can divide the area into triangles and rectangles to find the area. To calculate the area under a curve we have to count squares.

A Worked example

The speed–time graph for a train is shown below.

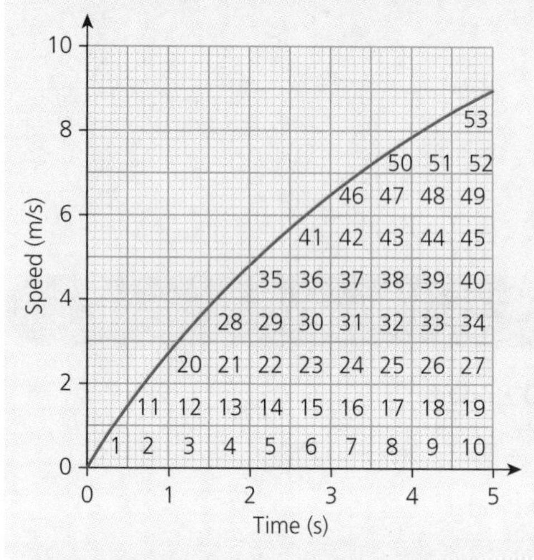

Estimate the distance travelled by the train in 5 seconds.

Step 1 The area between a speed–time graph and the time axis represents the distance travelled.

Step 2 To find this area, we count squares.

Step 3 Where the area is less than half a square, it is ignored. Where the area is more than half a square, it is regarded as a full square.

Step 4 Each full square represents a distance travelled of $1\,\text{m/s} \times 0.5\,\text{s} = 0.5\,\text{m}$.

Step 5 So, the 53 squares under this curve represents a distance of $53 \times 0.5 = 26.5\,\text{m}$.

Geometry and trigonometry

Geometry and trigonometry are areas of maths that look at angles, lines and shapes. This section looks at how these skills might be required in your science exams.

> **Key terms**
>
> Geometry: The branch of mathematics concerned with shapes and size.
>
> Trigonometry: The branch of mathematics concerned with the lengths and angles in triangles.

Using angular measures in degrees

Angles are measured in degrees. They can be any value between 0° and 360° (a full circle).

In your exam, it is likely that your knowledge of angles will be assessed in questions involving reflection and refraction. However, some exam boards may ask you to use a protractor to calculate a value.

You should know some of the more common values of angles in case you need to apply them to maths questions. For example:

- 90° is a right-angle
- 180° is a semi-circle
- 360° is a full circle – the angles of a segment in a pie chart therefore add up to 360° (see page 36)
- all angles in a triangle add up to 180°.

A Worked examples

1 The diagram shows rays of light as they pass from ethanol into the air.

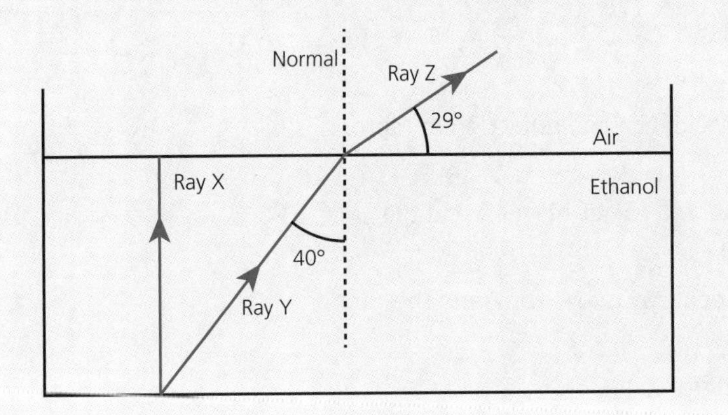

a Write down the angles of incidence of Ray X and Ray Y in ethanol.

Step 1 Ray X strikes the surface normally (at 90°), so the angle of incidence is 0°.

Step 2 Ray Y's angle of incidence is 40° (the angle between the incident ray and the normal).

b Calculate the angle of refraction of Ray Z in air.

Step 1 The angle of refraction is the angle between the refracted ray and the normal.

Step 2 This is 90° – 29° = 61°

2 A ray of red light passes through a triangular glass prism as shown. Calculate the angle of incidence in glass at surface B.

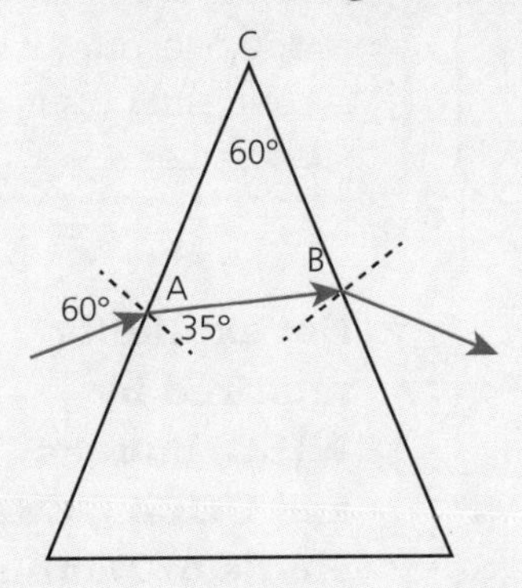

Step 1 At surface A, the angle between the ray AB and the internal surface of the glass is 90° – 35° = 55°.

Step 2 The angles in triangle ABC add up to 180°. So, at surface B, the angle between the ray AB and the glass is 180° – (55° + 60°) = 65°.

Step 3 The angle required lies between the ray AB and the normal at B. This angle is (90° – 65°) = 25°.

B Guided question

1 **Two mirrors are at right angles to each other. A ray of light falls incident on Mirror 1. The angle of incidence is 40°. The light eventually reflects off Mirror 2 as shown in the diagram.**

Calculate the angle of reflection at Mirror 2.

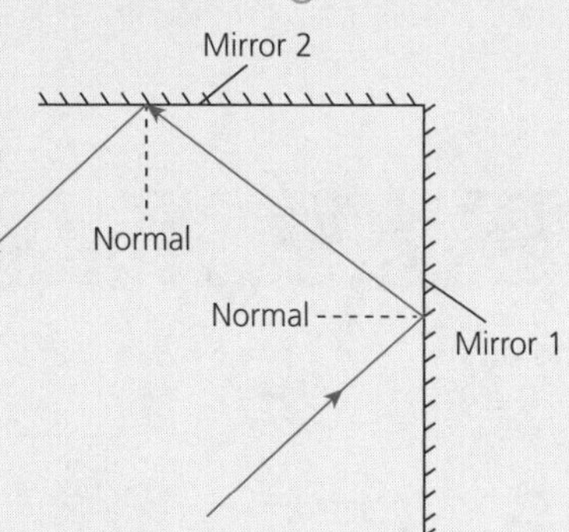

Step 1 Mark, on the diagram, the angles of incidence and reflection at Mirror 1, in degrees.

Step 2 Calculate the angle between the reflected ray at Mirror 1 and the mirror itself. Write the angle on the diagram.

Step 3 Calculate the angle between the incident ray on Mirror 2 and the mirror itself. (Hint: look at the triangle.)

Step 4 Calculate the angle of reflection at Mirror 2: ....

Key terms

Incident ray: A ray that strikes a surface.

Angle of reflection: Angle between a reflected ray and the normal.

Reflected ray: A ray that is reflected from a surface.

C Practice question

2 Two mirrors are inclined so that their reflecting surfaces are at 120° to each other. A ray of light falls incident on Mirror 1 with an angle of incidence of 70°. The light eventually reflects off Mirror 2.

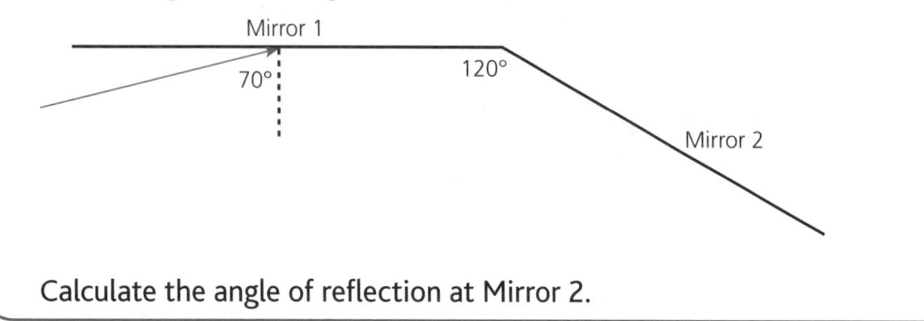

Calculate the angle of reflection at Mirror 2.

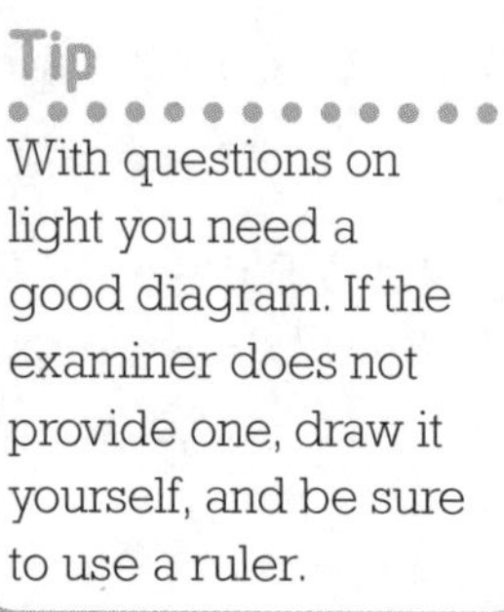

Tip

With questions on light you need a good diagram. If the examiner does not provide one, draw it yourself, and be sure to use a ruler.

Representing 2D and 3D forms

★ **Not explicitly required by WJEC/Eduqas and CCEA GCSE Single or Double Award.**

It is unlikely that you will be asked mathematical questions on this topic, but you do need to be able to represent 3-dimensional (3D) objects as 2-dimensional (2D) drawings.

Table 1.15 shows the differences between 2D and 3D shapes.

Table 1.15 Comparison of 2D and 3D shapes

3D shapes	2D shapes
Have 3 dimensions – length, depth and width.	Have 2 dimensions – length and width. They have no depth and are flat.
These figures can be drawn on a sheet of paper using wedged and dashed lines.	These figures can be drawn on a sheet of paper in one plane using solid lines.
3D figures deal with three coordinates, *x*-coordinate, *y*-coordinate and *z*-coordinate.	2D figures deal with two coordinates, *x*-coordinate and *y*-coordinate.

The displayed formulae and structural formulae represent molecules in 2D with all the covalent bonds shown as solid lines. No information about the orientation or shape of the molecules is given.

The displayed structural formulae of methane and ethanol are shown below.

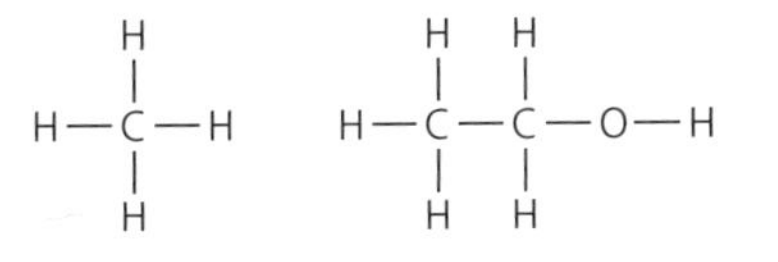

▲ Figure 1.20 Methane and ethanol

When drawing 2D structures it does not usually matter what angle you draw the atoms at. This is because molecules are three dimensional and when representing them in two dimensions on paper no diagram will represent them as they truly are. For example, each of the four structures in Figure 1.21 is a correct displayed formula of propene (C_3H_6).

▲ Figure 1.21

When molecules are drawn in 3D the symbols used are shown in Table 1.16.

Table 1.16 Types of bond

Type of bond	Orientation
Normal bond ———	Bond lies in the plane of the paper
Dashed bond ---------	Bond extends backwards effectively into the page
Wedged bond ▷	Bond extends forwards effectively out of the page

A 3D structural representation of methane, CH_4, with its ball and stick model is shown in Figure 1.22.

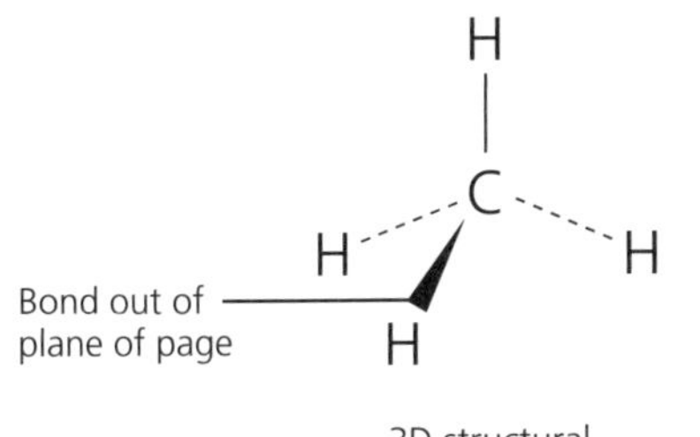

3D structural representation of methane

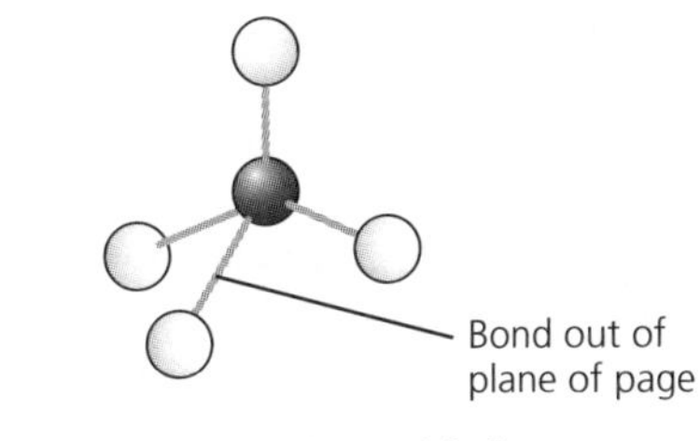

'Ball and stick' model of 3D structure of methane

▲ Figure 1.22 3D ball and stick model of methane (CH_4)

Graphite and diamond are giant covalent molecules. You should practise drawing diagrams of graphite and diamond.

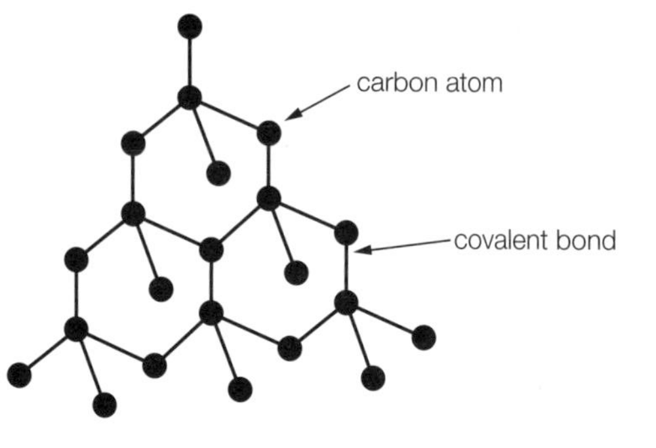

▲ Figure 1.23 Structure of diamond

▲ Figure 1.24 Structure of graphite

A Worked example

A molecule of ethanol has eight single covalent bonds. Draw the missing bonds in the diagram below to complete the displayed structural formula of ethanol.

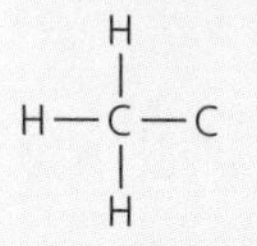

Step 1 Ethanol is C_2H_5OH. First fill in the bonds between the second carbon and two hydrogens as shown in red below.

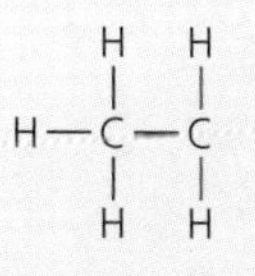

Step 2 Ethanol contains the OH group. Remember there is a bond between the O and the H that should be shown. Draw a bond between the carbon and the O and then another bond to the H (shown in red).

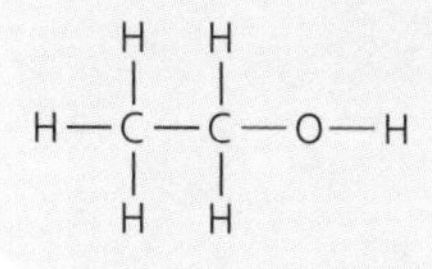

Tip

In Physics, you also need to know the symbols for cells, batteries, switches and lamps. You should not draw the components in 3D when creating a circuit diagram. For all sciences, you may also need to be able to draw simple diagrams of experimental apparatus, and so on.

B Guided question

1 **The diagram shows a 3D ball and stick model of a molecule of ammonia. Draw the 2D structure of ammonia.**

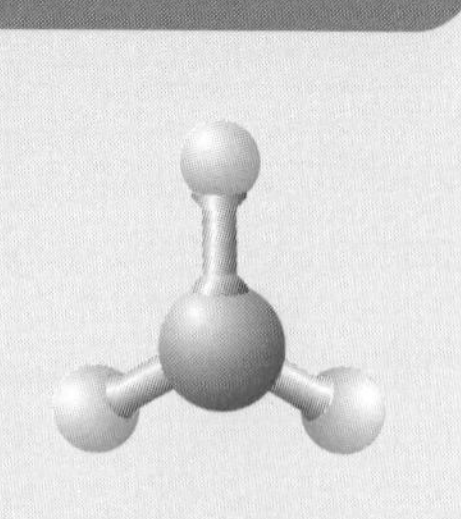

Step 1 You need to remember that ammonia has the formula NH_3 and so the red ball represents a nitrogen atom and the other three balls are hydrogen atoms.

Step 2 When drawing a 2D structure from a 3D representation remember that each stick represents a covalent bond, which is drawn as a single line. First draw the central nitrogen atom, as shown, then complete by drawing two more bonds to two hydrogen atoms.

N—H

C Practice questions

2 The diagram below shows a 3D model of a molecule of methane (CH_4). Draw the 2D structure of a methane molecule.

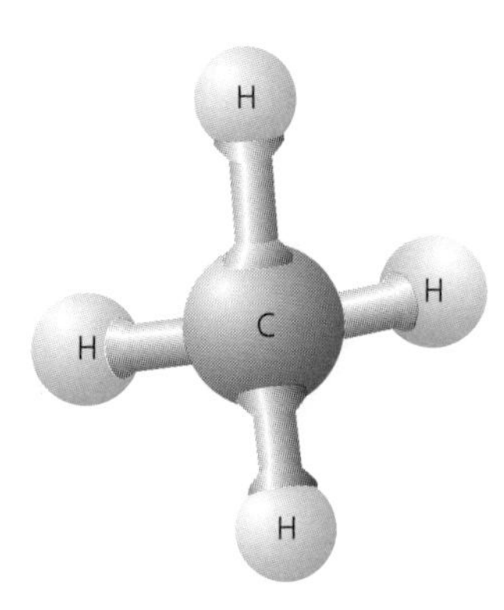

3 The figure below shows a dot and cross diagram of a molecule of water. Draw the 2D structure of water representing each covalent bond by a single line.

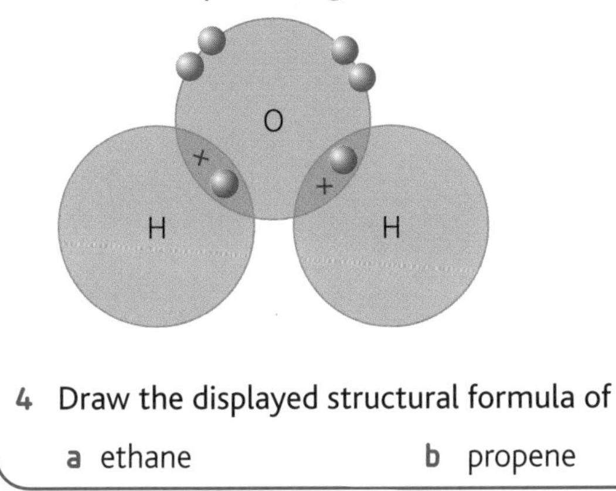

4 Draw the displayed structural formula of

a ethane   b propene   c but-1-ene.

Calculating areas and volumes

It is possible to calculate the area of triangles (see pages 71–76 for how this can be applied to graphs) and rectangles using the following formulae:

$$\text{Area of a triangle} = \frac{1}{2}\text{base} \times \text{height}$$

$$\text{Area of rectangle} = \text{length} \times \text{height}$$

The units of area are often mm^2, cm^2 or m^2.

> **Tip**
> The units for volume will be distance cubed (for example, in this case cm^3) while the units for surface area are cm^2.

You should also be able to calculate the surface area and volume of a cube. Surface area is the area of all the surfaces of a 3D shape added together. Volume is measured in cubed units such as mm^3, cm^3 or m^3.

To calculate the total surface area of a cube you need to calculate the area of each side and then multiply it by the number of faces.

If a cube has sides that are x cm long, the area of each side will be $x\,\text{cm} \times x\,\text{cm}$, or x^2.

A cube has six faces, therefore, its total surface area is $6x^2$.

To calculate the volume of a cube, you need to use the formula:
length × breadth × height

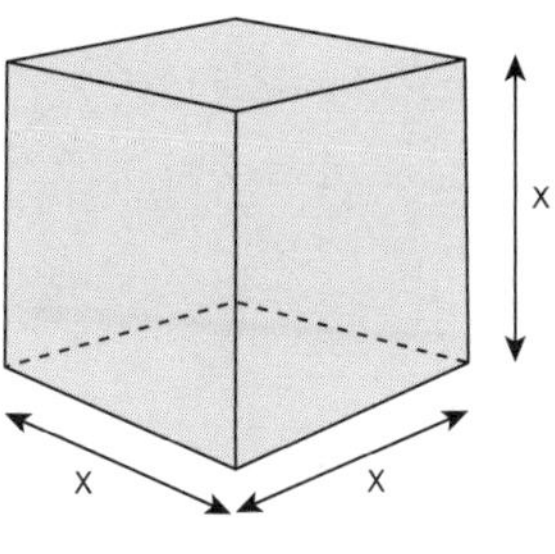

▲ Figure 1.25

As all the sides are the same length in a cube, if the side of a cube = x cm, this can also be represented as x^3.

Exam questions covering this skill may involve calculation of ratios, such as surface area to volume ratios. See the 'Ratios, fractions and percentages' section on pages 11–18 for more information on using ratios.

It is useful to work out the surface area to volume ratio, particularly to help understand rates of reaction. As the side of a cube decreases by a factor of 10 the surface area to volume ratio increases by a factor of 10.

A Worked example

a **A cube has side length 5 cm. Calculate the surface area and volume of this cube.**

Step 1 Calculate the surface area of one face of the cube.

Surface area of one face = $5 \times 5 = 25\,\text{cm}^2$

Step 2 Calculate the surface area by multiplying by the number of faces – there are 6 faces.

Surface area of one cube = area of one face × number of faces

$= 25 \times 6 = 150\,\text{cm}^2$

Step 3 Calculate the volume of the cube.

Volume of cube = length × width × height

$= 5 \times 5 \times 5 = 125\,\text{cm}^3$

b **What is the surface area to volume ratio of this cube?**

Step 1 Write down the ratio.

Surface area : volume

150 : 125

Step 2 Simplify the ratio by dividing by a factor such as 5. You may need to keep dividing both numbers by this factor until they can no longer be divided to give whole numbers.

Surface area : volume

150 : 125

30 : 25

6 : 5

The surface area to volume ratio is 6:5.

B Guided question

1 **Calculate the surface area : volume ratio of a cube that has a side length of 8 mm.**

Step 1 First calculate the surface area of the cube:

Surface area of cube = side length × side length × number of faces

Surface area of cube = ... × .... × 6

Surface area = ... mm^2

Step 2 Now calculate the volume of the cube:

Volume of cube = length × width × height

= .... × .... × ....

Volume = ... mm^3

Step 3 Finally, put these two values into a ratio: Surface area : volume = ...: ....

C Practice questions

2 In an investigation into diffusion, two different gelatine cubes were used. One had a side length of 6 cm and the other had a side length of 4 cm. Which of the cubes has the largest surface area to volume ratio? Show how you arrived at your answer.

3 Show that the surface area to volume ratio of a cube of side length 2 cm is ten times greater than that of a cube of side length 20 cm.

2 Literacy

Some questions on your GCSE Science papers will be extended response questions, which are usually worth six marks. As well as assessing your knowledge, these questions require you to construct a longer-form answer with a clear, logical structure. In other words, these questions also assess the quality of your written communication (QWC), sometimes known as quality of extended response (QER).

When answering extended response questions, you need to be sure your answer is:

- coherent – the points made in the answer are clear
- relevant – the points made in the answer all answer the question
- substantiated (for example backed up) – the points made in the answer are supported by scientific knowledge
- logically structured – the answer is well laid out, with points arranged in a logical order
- correctly punctuated, with technical terms spelt correctly.

This chapter will take you through the key points of answering these questions. The examples in this chapter all feature extended prose.

Tip

Most boards do not explicitly penalise you for inaccurate spelling, punctuation and grammar but if your language impedes understanding then it may cost you marks. You may not directly lose marks because of inaccurate spelling, punctuation and grammar, but if your answer is difficult to understand, then this may cost you marks.

» How to write extended responses

The first step in answering extended response questions well is to learn how to recognise them:

- Extended response questions will often use command words such as 'Evaluate', 'Explain', 'Design' and 'Compare'.
- These questions may require you to link knowledge, understanding and skills from more than one area of the specification, for example linking work on osmosis to the action of the hormone ADH.
- Extended response questions can also be multi-step calculations, although this is more common in the Chemistry or Physics papers.

Tip

Look at recent papers for your examination board to ensure that you can recognise the extended response questions.

Before starting your answer, it is useful to carefully read the question and ask yourself the following questions:

What is the question asking?

The most important part of an extended response question is identifying the command word – the key word that tells you what to do. Make sure your answer relates back to this command word, and answers the question asked.

Along with command words, extended response questions will often contain data and other key information. It is very important that you reference this data or information in your answer if provided – it is there for a reason. You may also occasionally see 'advice' in the question about what you need to cover to get top marks. Again, if this is the case, make sure you use it.

After reading the extended response question, the first thing you should do is underline the command word and think about what it means because it will tell you what the question is asking. After underlining the command word, read the question again and circle any words that tell you the topic that is being tested and any other key terms.

For example, to answer the question 'Compare the isotopes of lithium ^{6}Li and ^{7}Li', you would:

- Underline the command word 'Compare' and work out what this command word means. In this case, it wants you to describe similarities and differences.
- Then you should circle the word 'isotope' as this is the *topic* being tested; you would also circle the *key terms* '^{6}Li and ^{7}Li', which tell you which specific examples need to be included.

'Compare the isotopes of lithium ^{6}Li and ^{7}Li'

Having thought about what the question is asking, you should then plan your answer.

Key term

Command word: An instructional term that tells you what the question is asking you to do. 'Describe' and 'Explain' are two examples of command words.

Tip

Look for the command words in every question, regardless of how many marks it is worth. See pages 139–149 for more on command words and what they mean.

How do I plan my answer?

If you do not plan your answer, there is a temptation to write down everything you know about the topic, which means you will include many irrelevant details. Remember that the *quality* of your response and how well you answer the question is being assessed in this style of question, not how much you can write. It is, therefore, wise to quickly plan your answer rather than rushing straight into it.

First, think about the topic as a whole and decide what parts of your knowledge are relevant to the question. Then, consider how to structure your answer by putting the relevant points in a logical order.

For example, to answer the question on isotopes given above, the command word tells you that you need to 'Compare' – that is by describing similarities and differences – and to use the specific example of lithium given in the question. You could plan an answer in various ways, but some of the possibilities are outlined below:

1 Using a table

The command word is 'Compare' so a simple table showing the similarities and differences of the lithium isotopes may be useful.

Table 2.1

Same	^6Li	^7Li	Different	^6Li	^7Li
Number of protons	3	3	Number of neutrons	3	4
Atomic number	3	3	Mass number	6	7
Number of electrons	3	3			
Electronic configuration	2,1	2,1			
Reactions (same number of electrons in outer shell)					

You can then decide on the order you will cover your ideas – you could number the points in your table to help. Finally, when you write your answer, you can cross off each idea in the table once you have written about it.

2 Using a diagram

For this method, write down the key words you need to include and then add details about lithium as shown in Figure 2.1. Circle each key word and classify as a similarity or a difference.

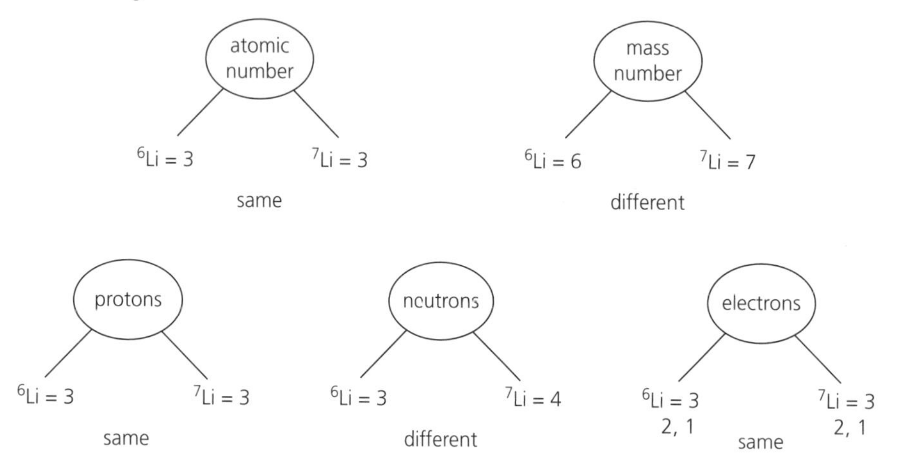

▲ Figure 2.1 Using a diagram to plan an extended response answer

Then, number the key words to show the order in which you will cover each idea. Finally, you should write your answer, crossing out each key word when you have written about it.

> **Tip**
> The examiner will mark everything you've written. If you don't want the examiner to mark your plans, remember to cross them out neatly or complete them on a separate scrap piece of paper.

3 Using bullet points

This method simply involves making short notes of what you will cover in your answer. The list format means it's easy to check through it quickly and you can also number the order you will cover each point. For example, for the isotopes question on page 85, you may write the following bullets:

- Things that are the same
  - both isotopes have same atomic number (3)
  - the same number of protons (3)
  - have one electron in the outer shell
  - and will react in the same way.

- Things that are different:
  - both isotopes have a different mass number (6 and 7)
  - a different number of neutrons (3 and 4).

When writing your answer, you will need to make sure these notes are rewritten as continuous prose, linking your ideas in a logical way.

Tip

In extended response questions you should not use bullet points, but complete sentences instead.

How do I check my answer?

For extended response questions you should read through your answer to check that

- you have answered what the command word in the question asked for
- you have used the examples or information you were asked to use
- your spelling, punctuation and grammar is correct.

In extended response questions correct spelling of key scientific words is important. There are a huge number of specific scientific words in an exam board specification, but it is a good idea to learn to spell the ones given below – they are those that are most commonly spelt incorrectly in extended response questions.

Table 2.2 Correct spellings of some important Biology words

mitosis	vaccination	gibberellins
meiosis	chlorosis	recessive
diffusion	aerobic	speciation
osmosis	anaerobic	archaea
capillaries	ciliary muscles	protist
communicable	pituitary	dialysis

Table 2.3 Correct spelling of some important Chemistry words

activation energy	alkali	burette
catalyst	collision	corrosion
covalent	crystal	delocalised
distillation	hydrochloric	molecule
nucleus	neutrons	neutralisation
pipette	polymerisation	phytomining
precipitate	solution	sulfuric

Table 2.4 Correct spelling of some important Physics words

acceleration	current	meter (the instrument)
angle	displacement	metre (the unit of length)
burette	fission	potential
centre	fusion	resistance
colour	gases	rheostat
convection	insulation	temperature
coulomb	longitudinal	transmission

Tip

In an examination, if you are asked to write an extended response answer, don't panic, just remember the steps using this acronym: CPWC – command word, plan, write, check.

How to do well in extended response questions

To fully understand how to successfully answer an extended response question it is useful to look at how such questions are marked – they are marked differently to other questions on an exam paper as you will be awarded marks according to the level of skill and knowledge that you show in your answer. This level is determined by:

- the overall quality of the answer
- the indicative content for each level.

You will note that mark schemes for these questions have three levels. This is why extended response questions are also called level of response questions.

When an examiner marks these question types, first they will read your answer as a whole and compare the quality of the content and writing with the level descriptor given on the mark scheme. They then decide which level best describes your answer – for example, if it is a strong answer they will give it a Level 3, while a weaker answer may get a Level 1. Have a quick look at pages 90 and 92 for examples of levelled mark schemes. To decide which mark within the level is given, the examiner then looks at the indicative content points, which are a guide to what should be included in the answer.

Indicative content is the factual points that could be included to answer the question. If the answer only barely meets the requirements the lower mark is awarded. You do not have to have all the indicative content points present to obtain full marks, but full marks are only awarded if there are no incorrect statements that contradict a correct response. The peer response style question on page 90 has an accompanying example mark scheme to help you further understand how this type of question is marked.

A general example of a six-mark extended response mark scheme is shown in Table 2.5:

Table 2.5 Example of a six-mark extended response mark scheme

Level	Description	Mark
Level 3	A clear, logical and coherent answer containing only relevant material.	5–6
Level 2	A partial answer with errors and some relevant material.	3–4
Level 1	One or two relevant points but lacks logical reasoning and contains errors.	1–2
	No relevant content.	0

Using the mark scheme above as an example:

- If a student's answer met all of the criteria for level 2 but not all of those for level 3, it would be placed in level 2. This could be due to the student including non-relevant information.
- If the student had written a good answer that only just missed out on level 3, they would be awarded 4 marks – the top of level 2.
- If, on the other hand, they had only achieved the very minimum to get into level 2, then they would be awarded 3 marks.

Tip

Remember you do not need to have all the indicative content points to obtain full marks, but your answer does need to be factually correct.

Tip

The key to answering extended response questions is to ensure that your answer contains all the elements that will ensure it is placed in the top band.

Tip

When revising, try and write an extended response answer that covers every one of these 'indicative content' points.

» How to answer different command words

Work through the following extended writing questions, which look at the main extended response command words. This will help you to further understand how to write a good longer-form answer.

For each command word there is:

- an 'expert commentary' question, which gives a sample student response, along with an analysis of what is good and bad about it
- a 'peer assessment' question where you will be given the chance to apply what you have learnt to mark a sample answer yourself
- an 'improve the answer' question where you will be asked to improve another student's response in an attempt to get full marks.

Extended responses: Describe

A question requiring you to 'describe' something is asking for a detailed written account on the relevant facts and features relating to the topic being examined. Remember that describe does not mean explain, which is a higher level command word. You do not need to focus on causes and reasons in describe questions.

A Expert commentary

1 Describe, in detail, how you would carry out an experiment to investigate how the resistance of a wire depends on its length. Your description should include details:

- of the circuit you need to set up
- about what you would do
- about what results you would record
- of how you would use your results to draw a conclusion. [6]

Student answer

Get a piece of wire. Connect it in a circuit with an ampmeter and voltmeter. Measure the amps and the volts. Divide the amps by the volts to get the resistence. Then repeat with another length of wire and so on until you have done it for six lengths. Then plot a graph. It will be a straight line. This tells us that the resistence is proportional to the wire.

This piece of work would probably score one mark only for the idea that the experiment requires the student to find voltage and current in order to measure resistance.

An examiner looking at this piece of work would immediately be surprised by its length. The question is worth six marks, but the student has only written a few lines of text and there is no detailed description.

There are also spelling mistakes in two scientific terms – the correct word is ammeter (and later the student spells resistance incorrectly; it should be resistance).

This is a mistake in describing how resistance is calculated – we need to divide the voltage by the current, not the other way around.

Finally, the student states the resistance is proportional to the wire. However, the question makes it clear that the experiment is to show how the resistance of the wire depends on its length.

It is unclear what circuit is to be constructed here.

It is unclear how the results are to be recorded.

The student is unclear what is plotted against what in the graph.

Tip

Bullet points in questions can be thought of as the scaffolding around which you must construct an answer.

B Peer assessment

2 Describe the process of in vitro fertilisation (IVF) treatment. [6]

Student answer

Eggs are collected from the mother. They are then mixed with some sperm from the father in the lab, and fertilisation takes place. These fertilised eggs develop into embryos, which are then grown in test tubes. At the start, the mother is given FSH, which are an enzyme, which stimulates the maturation of several eggs.

Use the mark scheme and indicative content below to award this answer a level and a mark.

Mark scheme

Level descriptor	Marks
Level 3: A clear, well-structured and logical answer where all the material is relevant. The answer clearly sets out the process of the IVF, including the use of FSH and LH, extraction of eggs, creation of embryos and implantation of the embryos into the mother.	5–6
Level 2: A reasonably clear and logical answer with some structure, where most of the material is relevant. Some parts of the process are not fully detailed.	3–4
Level 1: Few relevant points; a lack of clear structure or logical reasoning. The student gives a limited description of IVF which contains errors.	1–2
No relevant content.	0

Indicative content:
- The mother is given FSH and LH to stimulate the maturation of several eggs.
- The eggs are collected from the mother.
- The eggs are fertilised in the laboratory by sperm taken from the father.
- The fertilised eggs develop into embryos.
- Once the embryos are a small ball of cells, one or two are inserted into the mother's uterus.

I would give this a level of .... and a mark of ....

This is because

....

....

....

Tip

To assign a level, first look at the answer and decide if it is a logical series of steps that could be used to carry out IVF. Are there any significant steps missing?

C Improve the answer

3 Describe an experiment that could be used to electrolyse molten zinc chloride. In your answer name the apparatus you would use and state any observations. [6]

Student answer

I would weigh out 10 g of zinc chloride and record the mass. I would put it in an evaporating basin. I would put two electrodes in the zinc chloride and attach them to a power pack. When electricity is switched on at one electrode there should be a grey substance and at the other a gas would be observed. A fume cupboard should be used.

Rewrite this answer to improve it and obtain the full six marks.

Extended responses: Explain

'Explain' means to state the reasons why something happens. The points in the answer must be logically linked; for example, in this question the reason why each substance conducts should be given by naming the charge carriers.

A Expert commentary

1 Explain if the following substances conduct electricity by referring to the structures of the substances.

- copper metal
- copper chloride solution
- chlorine gas. [6]

Student answer

Copper is a metal and is used in wires and in plumbing in most households. It has lots of delocalised electrons which can move all around the layers and so it conducts electricity. This is why it is a good conductor. Copper chloride also has a metal in it so it has delocalised electrons. It cannot conduct when it is solid but when it is dissolved in solution then the delocalised electrons can move and carry the charge. Chlorine is a molecule and it does not have any plus or minus. Because it does not have a charge it cannot conduct electricity.

This is a Level 3 answer worth 5 marks.

This is extra information that is irrelevant, and it is a waste of time to include this. It may also lead to inaccuracies in your answer.

This is a correct description of how metals conduct.

Copper metal is not present in copper chloride so there are no delocalised electrons. Copper chloride is made up of copper ions and chloride ions, which move and carry charge.

The wrong particle for conduction is identified but the idea that particles can only move to conduct when in solution (or molten) is correct.

The student correctly recognises that charges are needed to conduct electricity, and chlorine has no charge.

B Peer assessment

2 Lipids are an important component of a balanced diet. Explain the importance of lipase and bile in the digestion of lipids. [6]

Student answer

Lipase is a digestive enzyme which breaks down lipids into amino acids. Bile is an alkali secretion which is stored in the gall bladder. It is released into the small intestine where it neutralises hydrochloric acid, which has been released from the stomach. Its main function is to emulsify lipids. This means causing the lipids to become small droplets, which increases the surface area. This speeds up digestion by lipase because the enzyme has a larger surface area on which to act. The alkaline conditions also increase the breakdown of lipids by lipase.

Use the following mark scheme and indicative content to award this answer a level and a mark.

Mark scheme

Level descriptor	Marks
Level 3: A clear, well-structured and logical answer where all the material is relevant. The answer clearly explains the importance of bile and lipase in the digestion of lipids with no key omissions or errors.	5–6
Level 2: A reasonably clear and logical answer with some structure, where most of the material is relevant. The importance of bile and lipase are both explained, but with omissions and some clear errors.	3–4
Level 1: Few relevant points; a lack of clear structure or logical reasoning. The answer is only a limited explanation of the importance of lipase or bile and contains obvious errors.	1–2
Level 0: No relevant content.	0
Indicative content: • Lipase is a digestive enzyme that breaks down lipids, producing glycerol and fatty acids. • The glycerol and fatty acids formed can then be used to produce new lipids. • Bile is an alkaline secretion that is produced in the liver, stored in the gall bladder and acts in the small intestine. • Bile neutralises stomach acid when it enters the duodenum. • Bile emulsifies fat to form small droplets, which increases the surface area of the fat. • The alkaline conditions and large surface area increase the rate of breakdown of the fat by lipase.	

I would give this a level of .... and a mark of ....

This is because

....

....

....

C Improve the answer

3 Use the definition of pressure to explain how the pressure due to a column of liquid in a measuring cylinder depends on the height h of the column. [6]

Student answer

Pressure is the force acting on a surface divided by the area of the surface.

The column of liquid is a prism of cross section area A and height h.

This means that the weight of liquid is $A \times h$.

So, the pressure depends on the height, because the weight depends on the height.

Rewrite this answer to improve it and obtain the full six marks.

Extended responses: Design, Plan or Outline

These command words are used in a similar way, so they have been grouped together here. In a 'Design', 'Plan' or 'Outline' question, you would be asked to describe how you would carry out an investigation or study.

You should not worry about giving exact volumes or masses as part of any method, but should ensure it is safe and appropriate to the context given. This means not including any equipment that would not be available, nor using an overly complex method.

Tip

The command words 'design', 'plan' and 'outline' are not exactly interchangeable, but they are often used in a similar fashion in GCSE Science. Namely, they will be asking you to set out an experiment to test a hypothesis.

A Expert commentary

1 A bacterial infection is proving resistant to common antibiotics such as penicillin. A pharmaceutical company would like to test the effect of the antibiotic tigecycline on the bacteria. Their hypothesis is that tigecycline will reduce the growth of the bacteria more than penicillin will.

Design an experiment, using discs soaked in antibiotics, to test this hypothesis. [6]

Student answer

Antibiotic solution of a known concentration and volume should be added to a series of discs. These discs should be placed on agar plates which contain a bacterial culture. The discs need to be placed in the centre of the agar plate. The plates then should be incubated at 37°C for 24 hours. At the end of this period, the clear area where no bacteria are growing should be measured.

This is a level 2 answer that would probably score three marks.

The details of the investigation are clearly laid out.

The end of the answer needs detail on comparing the results of the two antibiotics, and therefore concluding if the hypothesis is correct. The antibiotic that produces the largest clear area is the one that has reduced the growth of bacteria the most.

The key issue with this answer is that the hypothesis and the two different antibiotics are not referenced at all. It needs to be stated at the start of the answer that this method will be used for penicillin and tigecycline, and that the results can then be compared.

Tip

In school microbiology investigations, an incubation temperature of 25°C would normally be used to prevent the growth of human pathogens. As this is an investigation carried out by a pharmaceutical company on a human pathogen, 37°C is an appropriate temperature to use.

B Peer assessment

2 Outline an experiment to measure the angle of refraction in a rectangular glass block when the angle of incidence in air is 30°.

In your answer you must state the apparatus and method you will use. You should also draw a ray diagram to illustrate your plan and indicate the angles of incidence and refraction. [6]

Student answer

Place a rectangular glass block on a drawing board and draw around its outline with a pencil. Remove the block and draw the normal to one of the long sides.

Draw a line, L_1, at 30° to this normal at the point, P, were it meets the glass.

Replace the block and direct a ray of light along the line L_1 and observe the light exit the glass on the opposite side along line L_3.

Draw two small crosses on line L_3 and rule a line joining them back to the point Q were the light left the glass.

Remove the glass and rule a straight line L_2 between points P and Q.

Mesure the angel between line L_2 and the glass. This is the angel of refraction.

Diagram of apparatus:

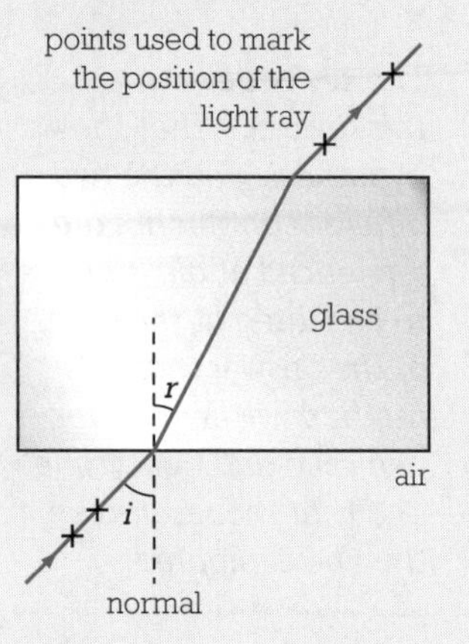

....

Use the mark scheme and indicative content below to award this answer a level and a mark.

Mark scheme

Level descriptor	Marks
Level 3: Detailed, well-structured plan that would work. At least six of the points in the indicative content are covered and the spelling, punctuation and grammar are largely accurate.	5–6
Level 2: The method may lack detail and structure, but with only minor changes it would work. At least four of the points in the indicative content are covered and the spelling, punctuation and grammar are usually accurate.	3–4
Level 1: The plan requires significant modification if it is to work. There may be significant irrelevant or incorrect information. At least two of the points in the indicative content are covered. There may be inaccuracies in spelling, punctuation and grammar.	1–2
No relevant content	0

Indicative content:

- set rectangular glass block on paper, outline with pencil
- use protractor to draw normal on one side
- draw line at 30° to normal at point of incidence
- direct ray of light from the ray box along this line
- method to trace emergent ray to point where light leaves the glass
- draw refracted ray in glass and measure angle between refracted ray in glass and normal
- diagram to show normal and angles of incidence in air and refraction in glass

I would give this a level of .... and a mark of ....

This is because

....

....

....

C Improve the answer

3 The order of reactivity of the halogens is shown below.

chlorine ↑

bromine — increase in reactivity

iodine

This order of reactivity can be determined by carrying out displacement reactions using aqueous solutions. Plan an experiment to determine the order of reactivity of these three halogens using displacement reactions. Include details of how the results can be used to determine the order. [6]

Student answer

First, place some potassium iodide solution in a test tube and add some aqueous chlorine. If there is a reaction, then chlorine is more reactive and displaces iodine from solution. In a second test tube you should place some potassium iodide solution and add some aqueous bromine. If there is a reaction, then bromine is more reactive than iodine.

Rewrite this answer to improve it and obtain the full six marks.

Extended responses: Justify

'Justify' and 'Evaluate' (see pages 143–144) are slightly different command words, but both require you to use evidence to make an argument. Both command words may be used together in the same question.

'Evaluate' requires you to use the information supplied in the question, as well as any relevant outside knowledge, to consider evidence for and against an argument. 'Justify', on the other hand, means you need to use the evidence supplied to support and take one argument forward. The main difference between these two words is, therefore, whether you have to assess arguments both for and against a conclusion, or simply the argument for a conclusion.

Tip

If numbers are supplied in an evaluate/justify question you are expected to use them. This means that your justification (or evaluation) may be mathematical in parts.

A Expert commentary

1 The table below details treatment strategies for measles. Justify these treatment strategies. [6]

Vaccination	All young children are vaccinated against measles.
Treatment	Antibiotics should not be used to treat measles. Patients should be given plenty of fluids and made to rest. Patients suffering from measles should be kept away from public areas.

Student answer

Vaccination is very important because it prevents people getting measles, which is a very damaging disease. Antibiotics should not be used because measles is a virus and antibiotics cannot be used to treat a viral disease. Fluids and rest are a better treatment than antibiotics for measles, and are the best treatment. The patients suffering from measles should be kept away from public areas as measles is very easily spread, and keeping people at home will reduce the spread.

The description of the use of the vaccination requires much greater detail.

The student gives a good justification for why antibiotics should not be used to treat measles.

More detail could be given on how measles is damaging, for example it can be fatal, and how measles is spread, as well as relating these ideas to the importance of keeping sufferers away from public areas.

This is a level 2 answer that would probably score three marks.

B Peer assessment

2 The volume of materials change with temperature. The graph shows how the volume of a fixed mass of liquid water changes as its temperature rises from 0 °C to 4 °C and then to 100 °C.

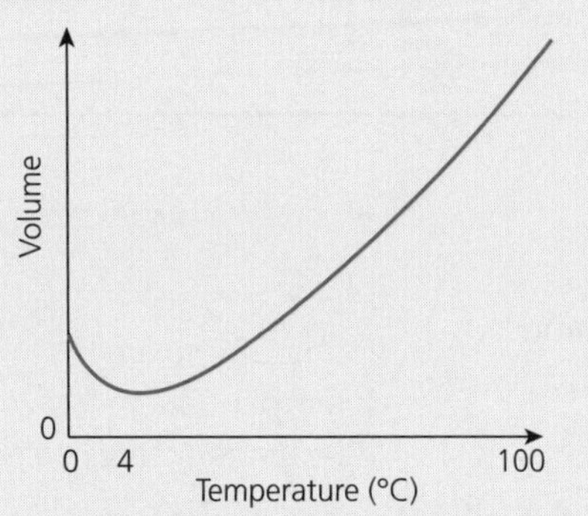

Study this graph showing changes in the volume of the liquid water over a given temperature range. Determine what is happening to the water's physical properties. Justify your reasoning.

In your answer you must refer to what is happening to the changes in:

- the separation of the molecules
- the density of the water
- the forces of attraction between water molecules. [6]

> **Tip**
> Always look carefully at graphs and identify their main features before starting your answer.

Student answer

1 As the temperature rises from 0 °C to 4 °C, the volume falls. This implies that the average separation of the water molecules is decreasing.

2 As the temperature rises from 4 °C to 100 °C the volume rises. This implies that the average separation of the water molecules is increasing.

3 At 100 °C the water changes into steam. Gases have lower densities than liquids, so the density must be decreasing.

4 At 0 °C the water changes into solid ice. Solids have higher densities than liquids, so the density must be increasing.

5 Between 0 °C and 4 °C the volume is falling, so the attractive forces pulling the molecules together are increasing.

6 Between 4 °C and 100 °C the volume is rising, so the attractive forces pulling the molecules together are decreasing.

Use the following mark scheme and indicative content to award this answer a level and a mark.

Mark scheme

Level descriptor	Marks
Level 3: Detailed, well-structured argument given. At least six of the points in the indicative content are covered and the spelling, punctuation and grammar are largely accurate.	5–6
Level 2: The argument may lack detail and structure. At least four of the points in the indicative content are covered and the spelling, punctuation and grammar are usually accurate.	3–4

Level descriptor	Marks
Level 1: The argument is broadly accurate. There may be significant irrelevant or incorrect information. At least two of the points in the indicative content are covered. There may be inaccuracies in spelling, punctuation and grammar.	1–2
No relevant content	0
Indicative content: The fall in volume from 0 °C to 4 °C implies: • a decrease in the separation of the molecules • an increase in the density, because the mass is constant • an increase in the attractive forces between molecules The rise in volume from 4 °C to 100 °C implies: • an increase in the separation of the molecules • a decrease in the density, because the mass is constant • a decrease in the attractive forces between molecules The water has: • maximum density at 4 °C	

I would give this a level of .... and a mark of ....

This is because

....

....

....

C Improve the answer

3 A farmer wanted to increase the yield of crops grown in a greenhouse. The graph below shows the effect of light intensity on the rate of photosynthesis. The farmer decided to also increase the temperature in the greenhouse, in addition to increasing the light intensity in the greenhouse.

Use the graph to justify the farmer's course of action. [6]

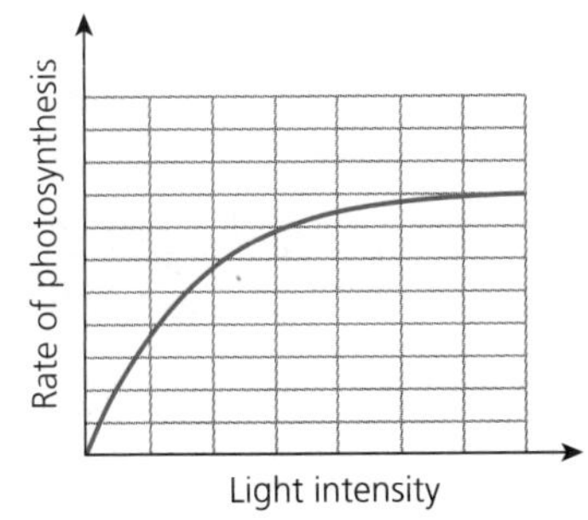

Student answer

Increasing light intensity increases the rate of photosynthesis. By increasing the light intensity in the greenhouse, the farmer will increase the amount of photosynthesis that their crops are doing, therefore they will grow more and the farmer will increase their yield. Temperature also effects the rate of photosynthesis, so by increasing the temperature the farmer will also increase the yield.

Rewrite the answer to this question so that it would be awarded all 6 marks.

Extended responses: Evaluate

In an 'Evaluate' question, you should use the information supplied and your own knowledge and understanding to consider the evidence for and against and draw conclusions.

Your answer is expected to go further than a compare question, as you need to give a final comment or deduction.

A Expert commentary

1 Diesel is the fuel used by most lorries. Research is being carried out into the use of hydrogen, instead of diesel, as a fuel for lorries. Evaluate the use of hydrogen rather than diesel as a fuel for lorries. [6]

Student answer

The raw material to make hydrogen is water and there is an abundant supply of water in the sea. Diesel comes from crude oil. When hydrogen burns it produces water only and so it does not cause any air pollution. Diesel however burns to produce carbon dioxide, which can cause the greenhouse effect. The greenhouse effect causes global warming and ice caps to melt. Incomplete combustion of diesel may produce carbon monoxide which is toxic and also carbon which can cause smog which causes respiratory problems. Hydrogen is produced from water, however as electricity is needed this also may produce pollution.

In conclusion, hydrogen is better to use because it is in good supply, and does not cause pollution but it is a flammable gas which is expensive to store safely.

This is a Level 3 answer, awarded 5 marks.

This clearly states a reason why hydrogen is good as a fuel.

A better evaluation here would be to continue to state that crude oil is in limited supply and is non-renewable.

The student has given good detailed evaluation of both fuels in terms of environmental problems caused.

More detail about the pollution produced by electricity generation would be useful.

This is good as it draws a conclusion that is consistent with the reasoning in the answer.

Tip

Always try to make a conclusion at the end of an evaluate question.

B Peer assessment

2 The graph below shows how the percentage yield of ammonia changes with pressure and temperature.

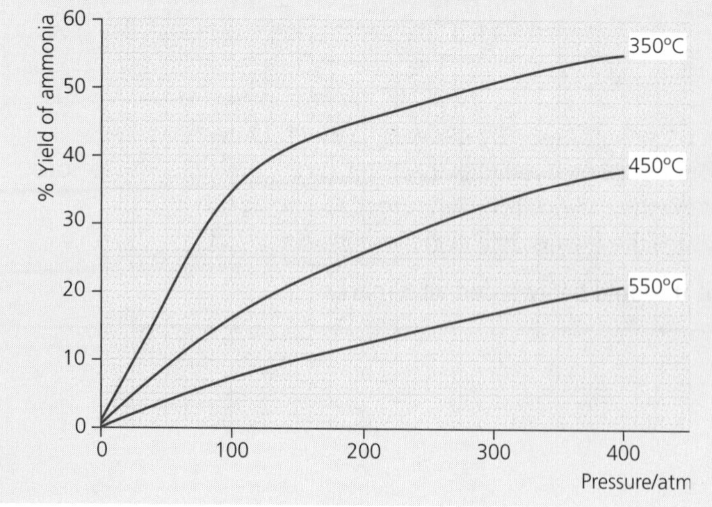

Use the graph and your own knowledge to evaluate the conditions used in industry to manufacture ammonia. [6]

Student answer

A low temperature gives a high yield of ammonia, but a low temperature makes this reaction very slow. Hence a compromise temperature should be used in industry which is moderate and gives a good rate and reasonable yield. Increasing pressure does not have much effect on the yield.

Use the mark scheme and indicative content below to award this answer a level and a mark.

Mark scheme

Level descriptor	Marks
Level 3: A detailed and coherent evaluation is provided that considers different conditions and comes to a conclusion for temperature and pressure consistent with the reasoning.	5–6
Level 2: An attempt to describe some conditions that comes to a conclusion. The logic may be inconsistent at times but builds towards a coherent argument.	3–4
Level 1: Simple statements made. The logic may be unclear and the conclusion, if present, may not be consistent with the reasoning.	1–2
No relevant content	0
Indicative content: • graph shows that increasing pressure increases yield • high pressure is expensive due to thick pipes • compromise pressure of 250 atm is used • this gives a reasonable yield at reasonable cost • graph shows that decreasing temperature increases yield • lower temperature decreases the rate • in industry a compromise of 450 °C is used • it is a compromise between a reasonable rate and reasonable yield	

I would give this a level of .... and a mark of ....

This is because

....

....

....

C Improve the answer

3 Chlorosis is a plant condition that can be caused by a lack of chlorophyll and proteins. A gardener was finding that large numbers of plants were suffering from this condition. They treated the plants by adding nitrate fertiliser.

Evaluate the suitability of this treatment for the plants. [6]

Student answer

This should be a suitable treatment for the plants because the condition is caused by a lack of proteins, and nitrates are used by plants for protein synthesis. By adding nitrates to the soil, the plant will be able to take these in and use them for protein synthesis.

Rewrite the answer to this question so that it would be awarded all 6 marks.

Extended responses: Use

The command word 'Use' means that the answer must be based on the information you are given in the question. Unless the information given in the question is used, no marks can be awarded. However, in some cases, you might also be asked to use your own knowledge and understanding.

> **Tip**
>
> 'Use' is another command word that will nearly always appear with other command words. Again, it is important to make sure you read the entire question properly.

A Expert commentary

1 The diagram shows some of the sound waves in a school assembly hall. Some of the audience complain that the sound they hear is not very clear.

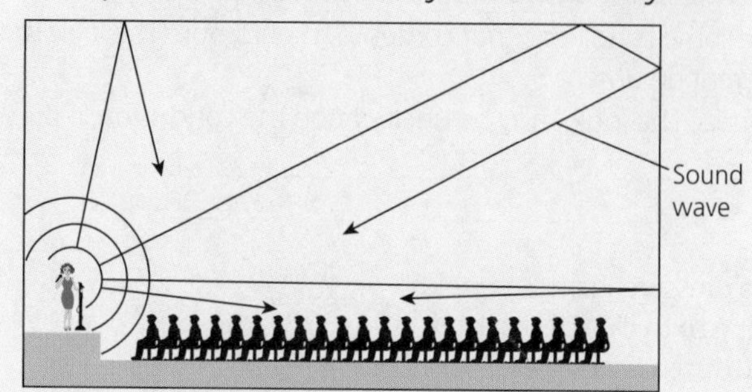

Use your knowledge of sound and the information provided in the diagram to explain why some people may not hear the sound clearly. You must refer to the cause of the problem, what the audience will hear and why, as well as how to correct the issue. [6]

Most of this answer is valid and creditworthy.

Most marks have been lost by supplying an incomplete answer. The student has not used the information in the diagram to point out that echoes are reflections, that the path lengths of the echoes are different and how the echoes can be reduced by using named, softer surfaces.

Student answer

Some people in the audience hear the same sound several times because of echoes. The echoes arrive at different times.

The student correctly identifies the cause of the problem as echoes arriving at different times and what has to be done to solve it.

Correcting the problem requires removing the echoes coming from the hard surfaces of the walls and sealing. This can be achieved by making the walls and sealing sound-proof.

There is only one spelling mistake. On two occasions the candidate referred to sealing instead of ceiling. This single mistake would be ignored.

The student may be a little confused about soundproofing. Soundproofing stops sound passing through a material and can be done using highly sound-reflective surfaces (which is the opposite to what is required here). They should instead be recommending sound absorbing surfaces to solve this problem.

This is a level 2 answer that would probably score three marks.

B Peer assessment

2 In an experiment, lithium bromide, sodium bromide and potassium bromide were dissolved in water. The temperature change observed for each solid was measured. The same amount of each compound and water were used each time.

Use the table of data to help you state and explain the results that should be obtained. [6]

Compound	Energy change when dissolving/kJ
Lithium bromide	−48.8
Sodium bromide	−0.8
Potassium bromide	+19.9

> **Tip**
>
> In a six mark question, there will be at least six ideas the examiner is looking for. You must provide at least six relevant points in your answer if you are to achieve full marks.

Student answer

For lithium bromide and sodium bromide the energy change is negative. A negative energy change means that the change is exothermic, heat is given out and the temperature has increased in the reaction. For potassium bromide the energy change is positive. This means that the temperature got colder in the reaction as heat is taken in. The results are for three group 1 metal compounds and they show a trend as you go down the group from lithium to potassium. The trend is that the temperature gets colder. Lithium bromide will show a greater temperature change than sodium bromide. It may also dissolve faster.

Use the mark scheme and indicative content below to award this answer a level and a mark.

Mark scheme

Level descriptor	Marks
Level 3: A detailed and coherent explanation is given that demonstrates a good knowledge and understanding and refers to energy changes and temperature changes for all three substances.	5–6
Level 2: An explanation is given that demonstrates a reasonable knowledge and understanding and refers to energy changes or temperature changes for the three substances.	3–4
Level 1: Some simple statements are made that refer to at least one energy change or temperature change.	1–2
No relevant content	0
Indicative content: • lithium bromide and sodium bromide give out energy/heat/are exothermic on dissolving • the temperature should increase when lithium bromide and sodium bromide dissolve • potassium bromide takes in energy/heat and is endothermic when dissolving • the temperature should decrease when potassium bromide dissolves • the energy given out when lithium bromide dissolves is much greater than when sodium bromide dissolves • the temperature change when lithium bromide dissolves is much greater than when sodium bromide dissolves	

I would give this a level of .... and a mark of ....

This is because

....

....

....

....

C Improve the answer

3 The diagram shows an ionisation smoke detector. The air gap between the source and the detector is about 1 cm wide. When there is no smoke, ionisation occurs in the air between the source and the detector. Provided enough ions arrive at the detector, no current is passed to the alarm circuit and the alarm is silent.

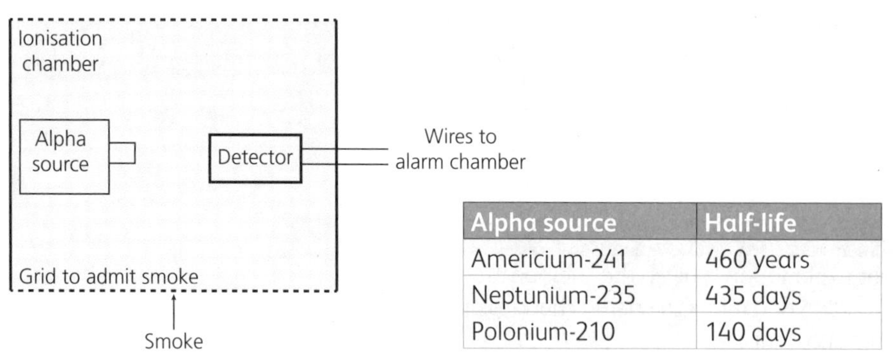

Alpha source	Half-life
Americium-241	460 years
Neptunium-235	435 days
Polonium-210	140 days

Use the information given above to describe how a smoke detector can set off an alarm. In your answer you must:

- state fully what is meant by ionisation in this context
- state how smoke coming between the alpha source and the alpha detector causes the alarm to ring
- state whether the smoke detector would work if the alpha source was replaced with a beta or gamma source
- use the information in the source table to suggest which source might be most suitable for a domestic smoke detector and why. [6]

Student answer

1. Ionisation occurs when an atom or molecule becomes charged. This occurs when it gains an electron and becomes negatively charged.
2. Smoke arriving in the ionisation chamber stops the alpha particles reaching the detector.
3. This causes a current to be sent to the alarm circuit and the alarm goes off.
4. The beta particles are too small to get into the detector and set off the alarm. The gamma rays are so penetrating that they would go straight through the detector and not set off the alarm.
5. The neptunium-235 is most suitable because 460 years for americium is too long in a domestic smoke detector. People would have died before it is used up. Polonium-210 is unsuitable because it is a dangerous poison.

Rewrite this answer to improve it and obtain the full six marks.

Tip

A question that asks you to give an answer 'in this context' requires you to apply your knowledge and understanding to the exact situation being discussed.

3 Working scientifically

Working scientifically is an area that is included as a required part of GCSE Science, although you will never be asked specific questions that are labelled 'working scientifically' in the exam. In reality, being able to work scientifically is a skill and a way of thinking – namely thinking like a scientist. This mindset can be hard to get into, but once you start thinking this way, it will be an incredibly useful skill for both GCSE and if you take science to A-level and beyond.

Most of the working scientifically skills will be covered as you work through your course. This chapter is all about making you aware of these skills, so that you can look out for where they feature in your studies, and use these opportunities to develop your thinking.

Working scientifically includes several different skills, which fall broadly into the following areas:

1. the development of scientific thinking
2. experimental skills and strategies
3. analysis and evaluation
4. vocabulary, units, symbols and nomenclature.

This chapter will deal with the first three areas, as vocabulary, units, symbols and nomenclature are covered in the maths and writing chapters of this book (see pages 1–83 and pages 84–102 respectively).

» Selecting apparatus and techniques

As part of your course, you are required to demonstrate your capability in using a range of apparatus and techniques (AT skills). These are techniques that you will develop over your course, probably as you complete the required practicals.

GCSE Biology AT skills

List of apparatus and techniques	
AT 1	Use of appropriate apparatus to make and record a range of measurements accurately, including length, area, mass, time, volume of liquids and gases, and pH.
AT 2	Safe use of appropriate heating devices and techniques, including use of a Bunsen burner and a water bath or electric heater.
AT 3	Use of appropriate apparatus and techniques for the observation and measurement of biological changes and/or processes.
AT 4	Safe and ethical use of living organisms (plants or animals) to measure physiological functions and responses to the environment.
AT 5	Measurement of rates of reaction by a variety of methods, including the production of gas, uptake of water and colour change of indicator.
AT 6	Application of appropriate sampling techniques to investigate the distribution and abundance of organisms in an ecosystem via direct use in the field.

List of apparatus and techniques	
AT 7	Use of appropriate apparatus, techniques and magnification – including microscopes – to make observations of biological specimens and produce labelled, scientific drawings.
AT 8	*(Single sciences only)* Use of appropriate techniques and qualitative reagents to identify biological molecules and processes in more complex and problem-solving contexts, including continuous sampling in an investigation.

GCSE Chemistry AT skills

List of apparatus and techniques	
AT 1	Use of appropriate apparatus to make and record a range of measurements accurately, including mass, time, temperature, and volume of liquids and gases.
AT 2	Safe use of appropriate heating devices and techniques, including use of a Bunsen burner and a water bath or electric heater.
AT 3	Use of appropriate apparatus and techniques for conducting and monitoring chemical reactions, including appropriate reagents and/or techniques for the measurement of pH in different situations.
AT 4	Safe use and careful handling of gases, liquids and solids, including careful mixing of reagents under controlled conditions, using appropriate apparatus to explore chemical changes and/or products.
AT 5	Safe use of a range of equipment to purify and/or separate chemical mixtures, including evaporation, filtration, crystallisation, chromatography and distillation.
AT 6	Making and recording of appropriate observations during chemical reactions, including changes in temperature and the measurement of rates of reaction by a variety of methods such as production of gas and colour change.
AT 7	Use of appropriate apparatus and techniques to draw, set up and use electrochemical cells for separation and production of elements and compounds.

GCSE Physics AT skills

List of apparatus and techniques	
AT 1	Use of appropriate apparatus to make and record a range of measurements accurately, including length, area, mass, time, volume and temperature. Use of such measurements to determine densities of solid and liquid objects.
AT 2	Use of appropriate apparatus to measure and observe the effects of forces including the extension of springs.
AT 3	Use of appropriate apparatus and techniques for measuring motion, including determination of speed and rate of change of speed (acceleration/deceleration).
AT 4	Making observations of waves in fluids and solids to identify the suitability of apparatus to measure speed/frequency/wavelength. Making observations of the effects of the interaction of electromagnetic waves with matter.
AT 5	Safe use of appropriate apparatus in a range of contexts to measure energy changes/transfers and associated values (e.g. work done).
AT 6	Use of appropriate apparatus to measure current, potential difference (voltage) and resistance, and to explore the characteristics of a variety of circuit elements.
AT 7	Use of circuit diagrams to construct and check series and parallel circuits including a variety of common circuit elements.

Look out for opportunities to tie these skills into your practical learning.

» The development of scientific thinking

This area of working scientifically is about understanding how scientific theories come to be developed and refined as well as recognising the importance of working safely and the limitations of what we can discover.

How theories develop over time

You may have heard of the scientific method before – this term is used to describe the process of formulating a hypothesis and then testing it by carrying out investigations. The results of these investigations can then be used to check the hypothesis, and either reject or refine it. Successful hypotheses can then be used to develop theories that explain natural phenomena.

Key terms

Scientific method: The formulation, testing and modification of hypotheses by systematic observation, measurement and experiment.

Hypothesis: A proposed explanation for a phenomenon used as a starting point for further testing.

You could be asked to give examples of how specific scientific methods and theories have developed over time. This could include how new data from experiments or observations have led to these developments. You could also be presented with some data, and asked if that data supports a particular theory.

A key example of this development is the theory of evolution over time, as outlined below:

Charles Darwin used his own observations, experimentation and the developing knowledge of geology and fossils to develop his theory of evolution. Darwin's theory was extremely controversial, and it was only as new evidence became available – including the mechanisms of inheritance – that it became widely accepted. The evidence helped disprove alternative theories such as those of other scientists like Jean-Baptiste Lamarck, who believed that changes during a single organism's lifetime could be inherited by their offspring. New discoveries, for example in the field of epigenetics, mean that our understanding of evolution will continue to develop.

Use scientific models

There are several types of model used to make predictions and develop explanations. The models you need to know are:

- Descriptive
- Mathematical
- Representational
- Spatial
- Computational

A *descriptive model* is like a picture you hold in your mind about the physical world. Examples include the particle model of atoms in liquids, or a description of the carbon cycle.

A *mathematical model* describes phenomena as a series of equations. You will frequently apply mathematical models in Physics. For example, all the equations you use to solve questions about electricity are based on the mathematical model of moving electrons.

A *representational model* represents something that we cannot see, with something we can see. For example, slinky springs are often used to show longitudinal and transverse waves. You know that sound is a longitudinal wave and light is a transverse wave although you have never *seen* the vibrations that cause them. Instead, the motion is represented by the motion in the slinky. Other examples include a model of the structure of a DNA molecule, or the use of dot and cross diagrams in Chemistry.

A *spatial model* is similar to both descriptive and representational models. For example, the Rutherford atom can be described in terms similar to our description of planetary motion and we can represent it on a two-dimensional diagram.

A *computational model* is a mathematical model run by a computer. For example, a computer model might be used to show the spread of an infectious disease in a population.

Tip

You could be asked to give the limitations of a particular model. All models have some limitations as they are always an imperfect representation of reality. The key to a successful model is one that is representative enough without being too complex.

Limitations of science and ethical issues

Science is an incredibly powerful tool to help us understand our world and also improve people's lives. However, it is only a tool, which means it has limits – both those imposed by the natural world and what is realistically achievable, and also limits that we impose ourselves.

Before a new scientific development is used, the factors listed below must be considered:

- cost
- the effect on the environment
- the effect on people
- ethics.

Key term

Ethics: This is the consideration of the moral right or wrong of an action.

When it comes to ethical concerns we need to constantly evaluate our use of science and decide whether any particular piece of scientific research is the 'right' thing to do.

One key ethical decision involves the use of living organisms in investigations. Exam questions may ask you to consider the ethical issues arising from a particular piece of research that involves killing the animal. In your answer to these types of questions, you might discuss ideas around an organism's 'right to life' balanced against the potential benefits of the research.

Other contentious ethical issues include genetically modified food, cloning, IVF, 'designer babies', or the use of stem cells. Many of these issues centre around the perceived rights of an embryo – some people believe that an embryo has a right to life, while other people believe that because an early embryo could not survive outside its mother, it is not truly alive, and therefore the benefits that derive from their use outweigh the ethical issues against their use.

Areas such as the above are often governed by law, which provides guidance on what is and isn't ethical (although some people argue that these laws limit what scientists are able to do). For new scientific developments in areas that outpace the speed of law, scientists have to take the difficult ethical decisions themselves.

Everyday and technological applications of science

You can see the impact of science all around you every day and everywhere you look, although the technological applications are the most obvious to us – it is difficult to imagine modern life without computers, engines or electricity. In the exam, you may be asked to describe examples of the technological applications of science within the specification.

Some of the example applications in the specification may include:

- treating coronary heart disease and heart failure, namely the use of stents to keep the coronary arteries open, and statins to reduce blood cholesterol levels
- vaccinations to reduce the spread of pathogens, and the importance of immunising a large proportion of the population

- detecting, identifying and treating plant diseases such as the tobacco mosaic virus and black spot
- the environmental implications of deforestation, global warming and a loss of habitats leading to a decrease in biodiversity
- fishing techniques, and how they can promote the recovery of fish stocks by controlling net size and introducing fishing quotas.

Exam questions on this area may ask you to evaluate methods that can be used to tackle the issues described in the question.

Tip

Details of how to answer 'Evaluate' questions can be found on pages 98-99.

Evaluate risks in science

Whenever scientists do practical work they need to evaluate risks – these risks could be minor and limited, or potentially catastrophic.

You will be familiar with the need to evaluate risk and should have carried out risk assessments when doing practical work. Risk assessments are an attempt to safeguard yourself, your colleagues and the apparatus. All experiments carry risk – even if it is only the risk that someone will, for example, swallow something.

A risk assessment requires you to:

1. think about *what* might reasonably go wrong
2. decide *how likely* it is to go wrong
3. consider *the effects* on people and equipment if the problem occurs.

Here is how to complete a risk assessment.

① Consider the physical hazards in the laboratory. For example, the things that you can trip over, electrical leads trailing from one bench to another, and so on. Make sure you attend to them before doing any practical activity.

② Identify specific hazards that relate to the experiment. These might be electrical (burns and shocks), radioactivity (exposure to radiation), high or low temperatures (burns and scalds), light (damage to eyes from lasers), and so on. These experiment-specific hazards are those that you need to identify in a GCSE exam.

③ Think about the effects of these hazards. They should be listed in order of importance.

- Death or permanent disability
- Long-term illness or serious injury
- Medical attention required
- First aid required

It is important that you are aware of all the possible risks and take action to avoid them. The greater the hazard, the more important it is that you take action to eliminate or reduce the risk.

In Chemistry, chemicals should have a COSHH hazard warning sign on the container. The ones you should recognise are shown below.

▲ Figure 3.1 Hazard warning symbols

Some examples of hazards, risks and control measures are shown in Table 3.1.

Table 3.1 Hazards, risks and control measures

Hazard	Risk	Safety precaution
Concentrated acid	Corrosive to eyes and skin	Wear safety glasses Wear gloves Use small amounts
Ethanol	Flammable	Keep away from Bunsen and flames Use a water bath or electric heater to heat
Bromine	Toxic	Wear gloves Wear safety glasses Use small amounts Use fume cupboard
Potassium/sodium	Explosive	Use small amounts Store under oil Wear safety glasses and use a screen
Cracked glassware	Could cause cuts	Check for cracks before use
Hot apparatus	Could cause burns	Allow to cool before touching Use tongs
Heating chemicals in test tubes	Chemical could spit out	Wear safety glasses Point test tube away from others
Long hair	Could catch fire	Tie back long hair

Tip

Wearing safety glasses is a requirement for *all* Chemistry practicals including all those in Table 3.1.

The importance of peer review

New scientific research is often published in scientific journals, but it is very unusual for respected publications to print research unless it has been peer reviewed. Peer review is a process by which scientific research is checked (and validated) by other scientists.

As part of this process, a paper will usually outline what the scientists were hoping to achieve, their methods, results and their conclusions. This paper will then be peer reviewed by independent experts to ensure that the research has been carried out correctly, that results are logical and that they support any conclusions drawn. This ensures that a piece of research is valid, and is vital to the development of scientific knowledge, particularly in achieving wider agreement and recognition, as well as identifying false claims.

In most cases, media representations of science are not peer reviewed, and can give a biased or inaccurate viewpoint which sometimes can lead to very real problems.

Key term

Peer review: The process by which experts in the same area of study evaluate the findings of another scientist before the research is considered for inclusion in a scientific publication.

Tip

Peer review can be seen on a small scale within your own class – sharing practical results with other students allows you to see if your results are consistent, and are therefore reproducible (see page 119 for details on reproducibility).

Questions

1. What is the scientific method?
2. How do scientists usually test a hypothesis?
3. Explain how the theory of evolution has developed over time.
4. Why are models important?
5. During IVF, some embryos are destroyed. Explain why some people may have objections to IVF.
6. How is the technical application of Biology helping to reduce the impact of overfishing?
7. A student is about to carry out an experiment to find the energy needed to change liquid water into steam.
   What would be the major hazard in this experiment?
8. A student wishes to use a Bunsen burner to heat a beaker of ethanol. Explain one suitable precaution, other than wearing eye protection, to reduce the risk of harm in this procedure.
9. A hypothesis for an experiment states that the solubility of potassium nitrate depends on the temperature of the water.
   a. Write down two general scientific actions that should be carried out after the hypothesis is written.
   b. Potassium nitrate is an irritant. Complete the risk assessment for the use of potassium nitrate in this practical.

Risk	Control measure
Potassium nitrate powder is an irritant	

10. Scientific research has shown a correlation between silver nanoparticles and the speed of healing of a cut. Explain how the scientific community would validate this research.

Experimental skills and strategies

As we have already seen, investigations are usually designed to test a hypothesis as part of the scientific method. The investigation is carried out, results are collected and the hypothesis is evaluated. This section covers the key areas needed to complete a successful investigation.

Developing hypotheses

Science is all about observations and asking questions.

For example, imagine you notice that a pendulum clock is losing time. You might first ask 'why is this happening?'. You guess that the time for the pendulum to make an oscillation (its period) depends on the weight at the end of the pendulum. This is your first hypothesis. To test this hypothesis, you carry out an experiment.

It turns out that your first idea is wrong. So, you put forward another idea – that the period depends on the pendulum's length. This is your second hypothesis. You carry out another experiment and find that you are right.

The important thing about a good hypothesis is that an experiment can be designed to test it – not whether it is correct.

In both the exam and in a required practical, you could be given some data and asked to suggest a hypothesis to explain the trend shown. Look carefully at the data, consider what part of your scientific knowledge it relates to and use this scientific knowledge to suggest the most likely hypothesis.

Tip

Remember that your hypothesis should not be too outlandish or surprising – make sure you sense-check what you have written. At GCSE, the questions will likely lead you to a fairly obvious answer.

Plan experiments to test hypotheses

In the exam, you could be asked to design or outline a practical procedure to test a particular hypothesis. To do this, you will need to use your own knowledge of the practicals you have completed and any information given in the question. You should also be able to explain why your chosen method is suitable for testing that specific hypothesis, and know why each of the steps needs to be carried out.

Tip

These types of questions usually contain the command words 'Design' or 'Plan'. See pages 92–95 for more detail on these command words.

When designing a practical investigation, you need to make sure that you are only changing one thing at a time as part of your tests. If you change more than one thing at once, it will be impossible to know which of the things you changed caused the results to be different, or even if both had an effect. The things that you can change are called 'variables'. There are three different types:

- Independent variable – this is the variable that is changed by the person doing the practical.
- Dependent variable – this is the variable that is measured during the investigation. We think that this variable is affected (or dependent) on changes in the independent variable.
- Control variables – these are variables that could affect the dependent variable. They need to be kept constant to ensure that it is only the independent variable causing any changes in the dependent variable.

Key terms

Independent variable: The variable selected to be changed by an investigator.

Dependent variable: The variable measured during an investigation.

Control variables: Variables other than the independent variable that could affect the dependent variable, and are therefore kept constant and unchanged.

Fair test: A test in which there is one independent variable, one dependent variable and all other variables are controlled.

Knowing what hypothesis you are testing for will affect which variable you decide to change, which you need to measure and which ones you must keep constant to make it a fair test. That is why knowing what each of the variables means will help you to plan experiments.

In the pendulum example above, the second hypothesis suggested that changing the length of the pendulum would change its period. We call the period the dependent variable, because it depends on something. We *think* it depends on the length of the pendulum – so the length of the pendulum is the independent variable.

Tip

You may also get asked to identify the different sorts of variable in an exam question.

Changing more than one thing each time might give misleading results. For example, if we think that the period might also depend on the mass of the pendulum, the mass must not change when we are testing the length. The mass is, therefore, a control variable.

In any science experiment there should only be *one* dependent variable and *one* independent variable. All the other variables must be controlled.

There are two other types of variable you need to know about. A continuous variable has values that are numbers. Mass, temperature and volume are examples of continuous variables. The variables used in Physics experiments are almost always continuous variables.

A categoric variable is one that is best described by words. Variables such as colour, shape and type of car are categoric.

Key terms

Continuous variables: The variables that can have any numerical value (such as mass, length).

Categoric variables: Variables that are not numeric (such as colour, shape).

Tip

For some practicals, especially in Biology, you may even have a controlled experiment to help you see the impact of changing an independent variable. For example, in identifying the conditions needed for photosynthesis, you might have a control plant where you would not change any variables as a point of comparison.

Select appropriate techniques, apparatus and materials

When completing practical experiments or questions, you may be asked what the best technique, instrument, apparatus or material would be for a particular purpose. In making your selection, you should be sure to know why it is the best choice.

Below are some questions to consider when making your choice of technique:

- Will this technique collect the data required by the investigation?
- Is the technique precise enough?
- Is it realistic to use this technique in this context?

For example:

- In investigations involving photosynthesis by aquatic plants, the rate of gas produced can be measured by counting bubbles produced in a certain amount of time. However, this is not a very precise measure of the volume of gas produced. A better technique would be to change the apparatus and use a gas syringe to give a much more precise measure of volume.
- When measuring very small volumes of liquid, a graduated pipette or syringe would be more appropriate than using a measuring cylinder.
- The use of complex equipment, such as laser measuring devices or electron microscopes, is unlikely to be realistic in the context of a question asking you to design an investigation to be carried out in a school lab.

In the pendulum example, you might consider the following:

You will need to measure length, time and weight. So, you need to choose the equipment to measure these quantities, as well as decide how best to use the equipment.

A metre stick is appropriate to measure the length as it is unlikely to be longer than this and you can see the length to approximately 1 mm. To use this metre stick correctly and ensure the test is fair, you have to measure the length in the same way each time. To do this you should ensure the object whose length is being measured is placed exactly alongside the metre stick and that both the object and metre stick are straight.

You need to consider whether there are any other obstacles to a fair measurement. For example, making sure there are no knots in the pendulum.

You must also make decisions about different techniques, such as what is the best length to measure. You could measure from the point of suspension to the bottom of the bob, or from the point of suspension to the middle of the bob. Again, you need to think scientifically. Weight acts from the centre of gravity, which is in the middle, so the best technique is to measure to the middle of the bob.

Measuring time is less accurate because there will always be an element of reaction time. To ensure that this effect is minimised, you need to think scientifically once more. You could pick a more accurate stopwatch (one capable of measuring to at least 0.1 s is probably suitable), or you could slow the pendulum swing by timing the period after allowing the pendulum to make a few swings first – starting the stopwatch when it reaches the end of a swing. It is also good practice to repeat the timing a few times and find the average period.

Some common apparatus that you might use are outlined in Table 3.2.

Table 3.2 Common apparatus

Apparatus	What the apparatus is used to measure
measuring cylinder / dropper pipette / syringe / graduated pipette	volume of liquid Note: Generally, a graduated pipette would be the most precise out of these pieces of apparatus, and the dropper pipette would be the least precise.
balance	mass
gas syringe	volume of gas
potometer	rate of water uptake by a plant
digital stopwatch	time
newtonmeter	force
protractor	angles
thermometer	temperature
ammeter	current
voltmeter	voltage

Carry out experiments appropriately and accurately

Planning is important in carrying out experiments appropriately. If you do not plan out an experiment properly, it can lead to results not being accurate or precise enough to draw reasoned conclusions. This is called an 'error in methodology', and is different from carrying out the experimental techniques incorrectly.

To ensure you avoid these types of errors, you need to think about potential issues in specific experiments. For example:

- If using a thermostatically-controlled water bath, ensure that there is time for the sample in the water bath to reach the same temperature as the water bath. Students often put test tubes into a water bath and begin collecting results immediately. This is incorrect as the contents of the test tube will take time to reach the desired temperature.
- When carrying out an investigation involving living organisms, you should allow any organisms to acclimatise to their surroundings before beginning to take measurements. This is because organisms will often be stressed by being moved or put into new surroundings, and this may affect the variable you are attempting to measure.

It is also really important to ensure that the equipment you select is accurate enough to gather the data required, for example, if the differences in mass between two samples are likely going to be very small, you should ensure the balance you are using is precise enough to distinguish this small change. When picking your equipment, it is useful to think about the differences between accuracy, reliability, precision and resolution.

Accuracy

Accuracy is how close we are to the true value of a measurement. For example, imagine if five students measured the growth of the same plant over a certain period in normal conditions. They got the following results: 15 mm, 17 mm, 16 mm, 17 mm and 12 mm. We know there can only be one true value as they have all measured the growth of the same plant.

The mean (see page 26) is probably our best (most accurate) value for the total growth, because it is likely that some of the students got too high a value and some got too low a value. The mean lets us balance out the values that are too high with those that are too low. In this example, we might discount the 12 mm result, which seems like an outlier, to give a mean of 16 mm. Remember that to improve accuracy, you should repeat and then average.

Another way to improve accuracy is to use a better measuring instrument. A good digital voltmeter, for example, is likely to be more accurate than an inexpensive analogue meter.

Resolution

Resolution is the fineness to which an instrument can be read. For example, a stopwatch with a sweeping hand might have a resolution of $\frac{1}{10}$ of a second, while a digital stopwatch might have a resolution of $\frac{1}{100}$ of a second. However, both stopwatches have the same precision because this factor will be determined by the reaction time of the person using it.

Precision

Precise measurements are those where the range is small. For example, suppose three students testing energy content within food measured the temperature change on one thermometer labelled 'A' and got the results: 3°C, 7°C and 6°C. The range of these measurements is 7 − 3 = 4°C, and the mean is 5°C.

Suppose these students had carried out the exact same experiment with another thermometer labelled 'B' and had got the following results: 4°C, 6°C and 6°C. The range of these measurements is 6 − 4 = 2°C, but the mean is still 5°C, as with thermometer 'A'. The readings are equally accurate, but those taken with thermometer 'B' have greater precision.

Reliability

A test is defined as reliable if different scientists repeating the same experiment consistently get the same results. The technique to improve reliability is to repeat the same test several times.

Tip

It can be quite easy to get these terms confused, so make sure you are clear on their definitions and how they are different.

Key terms

Accuracy: How close we are to the true value of a measurement.

Reliability: An experiment is reliable if different people repeat the same experiment and get the same results.

Precision: Precise measurements are those where the range is small.

Resolution: The fineness to which an instrument can be read.

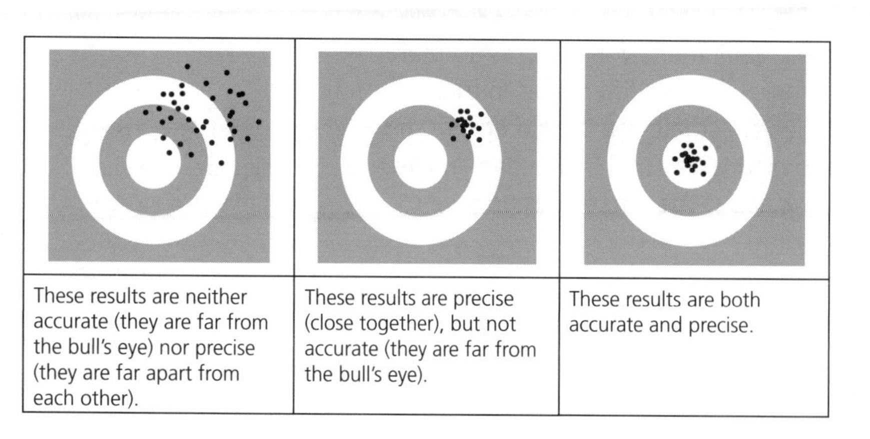

▲ Figure 3.2 Diagram showing the meaning of accuracy and precision

Sampling techniques

When gathering sample data, it is important to ensure that the data is representative. This means that the sample data collected is typical of the overall area being sampled.

> **Key term**
> Representative data: Sample data that is typical of the overall area or population being sampled.

This skill is especially important during ecological sampling. When investigating abundance of a plant in a field, you will be sampling a small area of the field and then using your sample results to estimate total abundance in the field. It is therefore very important that the area you are sampling is representative of the rest of the field. Below is a possible method that should produce representative results:

- Select a sample area, for example 10 m by 10 m.
- Use a random number generator to generate co-ordinates of where to place your quadrats.
- Sample at least 10 quadrats in the sample area.
- Repeat the process in other sample areas in the field.

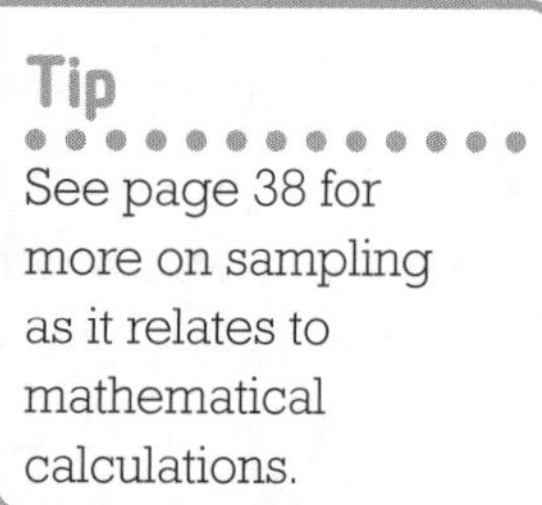

Taking multiple random samples in each sample area and using more than one sample area increases the chance of your results being representative of the field as a whole.

Make and record observations

This is an important skill when carrying out practical activities, and it is important that you make and record measurements carefully, as well as double-checking them to ensure you have not made an error when reading off a scale.

> **Tip**
> A parallax error can cause you to misread a scale. To avoid this type of error, make sure your eye is level with the measurement apparatus so that your line of sight is consistent.

Making observations

You need to make sure you plan to collect data in a timely fashion. If a reaction is occurring quickly, collecting data every five minutes might be inappropriate as the changes you are attempting to measure are occurring too fast. Similarly, if a process is occurring quite slowly, measuring every thirty seconds may be an inefficient use of time, and generate lots of data points that are not useful.

> **Key term**
> Parallax error: A difference in the apparent value or position of an object caused by different lines of sight.

Some other common issues with making observations and measurements to look out for include:

- Failing to note the measurement correctly – either through misunderstanding what the scale is showing or through misreading it. To avoid this, make sure you are clear on what the scale is showing, including what the minor graduations represent, for example what the lines between 10 cm^3 and 20 cm^3 on a measuring cylinder represent.
- Not using a stopwatch correctly – ensure you are confident using the stopwatch you have been given – including starting, stopping and clearing it. Make sure you are able to start the stopwatch at the appropriate time and finish it precisely at the correct end point.
- Not zeroing a balance before measuring mass – ensure the balance reads zero before measuring a mass. This can involve placing a container on the balance, zeroing it and then placing the sample into the container. Failing to do this can mean that the mass you measure is the sample as well as the container it is being weighed in.
- Not accurately determining a colour change – measuring colour change is generally subjective due to the fact that different people may judge the final end point slightly differently. To help with this, you should use a reference sample that has already reached the end point of the colour change.

Tip

Colour is a type of categoric variable.

Recording observations

During experimental activities you will usually record results in a table. When drawing tables ensure that:

- The table is a ruled box with ruled columns and rows.
- There are headings for each column and/or row.
- There are units for each column and/or row – usually placed after the heading after a solidus (/) or in parentheses ( ) for example 'Temperature /°C' or 'mass (g)'. Units should not be written in the body of the table.
- There is room for repeat measurements and averages – remember the more repeats you do the more reliable the data.
- The independent variable is recorded in the first column, and the dependent variable can be recorded in the next columns.
- Data should be recorded to the same number of decimal places or significant figures.

Qualitative observations are what we see and smell during reactions. Important types of observations in experiments, and notes on how to record these, are shown in Table 3.3.

Table 3.3 Types of observation

Type of observation	Notes on recording observations	Examples
Colour change	Always state the colour of the solution before the reaction and after.	*When bubbling an alkene into bromine water – the colour change is an orange solution to a colourless solution.*
Bubbles produced	If a gas is produced, then bubbles are often observed in the liquid and the solid reactant disappears.	*When sodium carbonate reacts with an acid, the observation is bubbles and the solid reactant disappears.* (Note, writing 'carbon dioxide is formed' is **not** an observation.)

Type of observation	Notes on recording observations	Examples
Heat produced	In many reactions the temperature changes.	*When acids react with alkalis, the temperature increases.*
Precipitate produced	When two solutions mix, often an insoluble precipitate forms. Ensure you use the word 'precipitate' in your observation as a common mistake is to write that the solution becomes cloudy. Also, do state the colour of the precipitate and the colour of the solution before adding the reagent.	*When barium chloride solution is added to a solution containing sulfate ions, a white precipitate is formed in the colourless solution.*
Solubility of solids	When a spatula of a soluble solid is added to water, the observation is often that the solid dissolves to form a solution. Make sure you state the colour of the solution formed.	*Copper(II) sulfate crystals dissolve in water to produce a blue solution.*
Solubility of liquids	When a liquid is added to water, always record if it is miscible or immiscible with water.	*Ethanol and water are miscible.*

Tip

Remember that clear is not a colour – instead use the word 'colourless'. For example, hydrochloric acid is *colourless*. It is also clear, but this refers to the fact that it is transparent.

Evaluate methods and suggest possible improvements

Evaluating means assessing how it is going as you carry it out, and, at the end, thinking about what could be improved if it was to be carried out again. You should always be evaluating while carrying out practical work.

Key term

Evaluate: This means to weigh up the good points and the bad points.

Evaluating is an important part of the scientific method. Written evaluation is often required as a part of a conclusion.

Tip

See pages 98–99 for further examples of 'Evaluate' questions.

During an experiment, you may find that some ideas, apparatus or techniques are not working well. If you need to make changes to improve your method, do so and refer to it in your evaluation. Scientists have to be flexible – but you need to explain *why* you changed your plan. Remember that you will need to repeat all your tests whenever you change your approach to ensure it is fair.

Evaluating may involve assessing whether:

- sufficiently precise measurements have been taken in an experiment – if an experimental method lacks precision, it may produce results that are not valid
- the method used in the investigation could be improved – you should be able to justify your answer and suggest improvements to ensure the results of an investigation are valid.

When answering evaluation questions, you should consider the following:

- Did the measurements taken provide data that can be used to answer the hypothesis that the investigation set out to address convincingly?
- Are there any weaknesses in the experimental method or the conclusion drawn from the results?
- Are there any improvements that could be made to the experimental method that would have produced more accurate or precise data?
- Will any follow-up experiments be needed to address any further issues raised by the original investigation?

Questions

1 Identify the independent variable, dependent variable and one controlled variable for each of the following investigations.
  a An experiment is carried out to see if there is any link between the resistance of a piece of wire and its length.
  b A student is investigating Newton's Second Law. The student wants to find out the relationship between the applied force on a trolley and its acceleration.
  c In the reaction between copper carbonate and hydrochloric acid, the time taken for a mass of copper carbonate to be completely used up was recorded. The experiment was repeated using different masses of copper carbonate.
  d The volume of carbon dioxide gas produced when calcium carbonate reacted with hydrochloric acid was measured, and the experiment repeated using different masses of calcium carbonate.
  e An investigation was carried out into effect of light intensity on the rate of photosynthesis of a sample of pondweed. A light was placed at a range of distances from the pondweed, and at each distance, the number of bubbles released from the pondweed in 5 minutes was counted. The same species and mass of pondweed was used throughout the investigation.

2 In an investigation 2 g of copper(II) sulfate crystals were found to dissolve faster in hot water than in cold water.
  a Write a hypothesis for this investigation.
  b State an observation that occurred in the investigation.
  c Name three pieces of apparatus that would be used in this investigation.

3 Copy and complete the diagram to show how you can distil copper(II) sulfate solution and collect pure water. Label the pure water and the copper(II) sulfate solution.

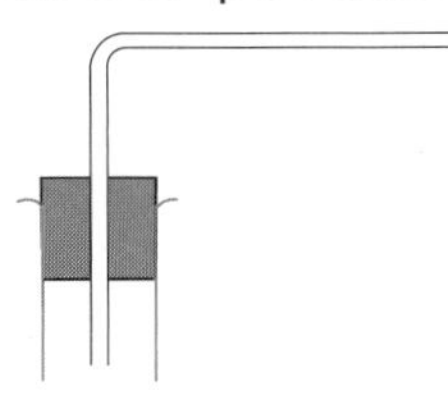

4 State the most suitable apparatus for measuring each of the following quantities.
  a mass of about 25 grams to a precision of 0.1 gram
  b volume of 24.7 cm^3 of liquid
  c time of about 45 s to the nearest 0.1 s
  d current of about 0.050 A to the nearest 0.001 A

5 How do errors in methodology differ from errors made while carrying out an investigation?

6 If investigating the abundance of a plant in a field, why is it important to take a number of different samples?

7 In an experiment a student weighed some hydrated magnesium sulfate crystals, heated them for 2 minutes and reweighed. The results recorded are shown in the table below.

Mass of crucible and hydrated magnesium sulfate before heating /g	Mass of crucible and magnesium sulfate after heating /g
9.37	8.25

  a What is the resolution of the balance used by the student?
  b Suggest one improvement that could be made to the results table.

8 A Physics student measures out five different volumes of ethanol and measures the mass of each. The results obtained are:
20 cm^3 16 g; 35 cm^3 28 g; 45 cm^3 36 g; 50 cm^3 40 g; 55 cm^3 42 g
  a Present these results in a suitable table with headings.
  b Which result might the student want to repeat? Why?

9 50 cm^3 of hydrogen peroxide and 1.0 g of manganese dioxide were allowed to react at 25 °C. The volume of oxygen collected from the reaction at 10 second intervals is: 8 cm^3 after 10 seconds, 30 cm^3 after 20 seconds, 49 cm^3 after 30 seconds, 59 cm^3 after 40 seconds and 63 cm^3 after 50 seconds. On repeating the experiment, the volume of gas obtained at each time interval was 32, 51, 59, 63, 65 cm^3 respectively.
Present these results in a suitable table with headings and units. Calculate and record the average volume of gas produced.

10 Three students are designing an experiment to measure the time taken for a trolley to roll down a ramp. They consider three timing methods:
- using a stopclock capable of measuring time to the nearest second
- using a stopwatch capable of measuring time to $\frac{1}{100}$th of a second
- using a light gate and datalogger, capable of measuring time to $\frac{1}{100}$th of a second

The true time to run down the ramp is 9.5 s.

a Which instrument has the least resolution?

b The students all decide to use the datalogger method and, independently, they obtain the following times (to 1 d.p.):

Student	Time (seconds)				
A	9.8	9.3	9.9	10.3	10.3
B	9.8	9.8	9.8	9.9	9.8
C	9.5	9.4	9.6	9.5	9.5

i Which student's results have the greatest precision?

ii Which student's results are neither reliable nor precise?

c Student A repeats the experiment. Why is this a good idea?

» Analysis and evaluation

Once experimental results have been collected, they need to be presented, analysed and then evaluated so that you can write a reasoned conclusion.

Collecting, presenting and analysing data

When carrying out experiments you will often record your results in a table. However, it can be difficult to spot trends from a table, so you may decide to create a graph. Once in graph form, the gradient of the line and its intercept on the vertical axis can provide a more obvious indication of a trend. For more details on graphs, tables, distributions and analysing data, see the Maths skills chapter of this book.

Evaluating data

Evaluating the quality of data obtained in an investigation is very important because poor quality data may mean that any conclusions drawn from it are incorrect. When evaluating the quality of data collected in an investigation, you can talk about accuracy, precision, repeatability and reproducibility, but you should also consider uncertainty and errors that may have been made in the experiment.

Uncertainty

All data collected by a scientist contains an element of uncertainty. This can be due to the instrument's lack of precision or inconsistencies in measurements made by the individual.

Uncertainty in a measurement is the maximum difference between the mean value and the experimental values.

If a student makes three measurements of the density of a liquid and obtains values (in g/cm^3) of 1.48, 1.53 and 1.49, the mean is 1.50 g/cm^3, and the uncertainty is calculated as $1.53 - 1.50 = 0.03\ g/cm^3$. A large uncertainty means poor precision. Uncertainty can be represented on a graph by range bars. The larger the range bars, the more uncertain the results.

This is a problem because we cannot tell which of the values taken is best for our measurement.

The cause of experimental uncertainty is known as error. There are two types of error – random error and systematic error. Make sure you are aware of them so you can take steps to eliminate or reduce their effects.

Key terms

Random error: An error that causes a measurement to differ from the true value by different amounts each time.

Systematic error: An error that causes a measurement to differ from the true value by the same amount each time.

Random error

A random error is one that causes a measurement to differ from the true value by different amounts each time. Three students measuring the volume of a cube are likely to come up with three different values. This is the result of random error. The error is randomly scattered about the true value. By making more measurements and calculating a new mean, we reduce the effects of random error.

Systematic error

A systematic error is one that causes a measurement to differ from the true value by the same amount each time. This will usually be due to the equipment used. If you have ever plotted a graph and found that it crossed the vertical axis when you expected it to go through the origin, the cause is likely to be a systematic error as *all* the results are different by the same amount. Systematic errors cannot be dealt with by simple repeats. Instead, you need to use a different technique or apparatus.

Repeatability, reproducibility and validity

When analysing results, they should also be ideally repeatable, reproducible and valid to ensure that they are useful.

- Results are repeatable if similar results are obtained when the investigation is repeated under the same conditions by the same investigator.
- Results are reproducible if similar results are obtained by different investigators using different equipment.
- Results are considered valid if the data is a correct measure of the property being investigated. For example, measuring the loudness of a sound is not a valid way to measure its speed.

We call results reliable if they fulfil the conditions of being valid, repeatable and reproducible. Reliable results are important in assessing whether we have discovered something meaningful. If results were, for example, repeatable but not reproducible, or reproducible but not valid, the results may be incorrect. Results that are simply repeatable are particularly untrustworthy as the person carrying out the experiments may just be repeating their mistakes again and again. Reproducible results give greater confidence that they are correct because several people or techniques are involved, but if the wrong thing is being investigated, it is still not valid or meaningful.

Tip

Anomalous results are values that are very different to the rest of the results from an investigation. If they have been produced by incorrect measurement, then they can be ignored when calculating means or carrying out further analysis.

Tip

Measurements can be repeatable but still subject to errors caused by the equipment used or by the investigator's experimental technique. Reproducible results are less likely to contain such errors because the results are gathered using different equipment and by different investigators.

Questions

1 What are anomalous results?

2 Two students carried out experiments to find the percentage by mass of nitrogen in ammonium sulfate. The true value is 21.2%.

Student A	21.4%	21.2%	21.1%	21.3%	21.4%
Student B	22.5%	22.4%	22.6%	22.5%	22.5%

a Calculate the mean value for each student. State the uncertainty in the mean.
b Comment on the accuracy of the mean result for each student.
c Comment on the repeatability of the result for each student.
d Why did the results differ when each student repeated them?
e Explain if either of the students had a systematic error in their experiment.

3 In an experiment some calcium carbonate and acid were placed in a conical flask on a balance and the balance reading recorded every minute. The results were recorded and the graph shown below was drawn.

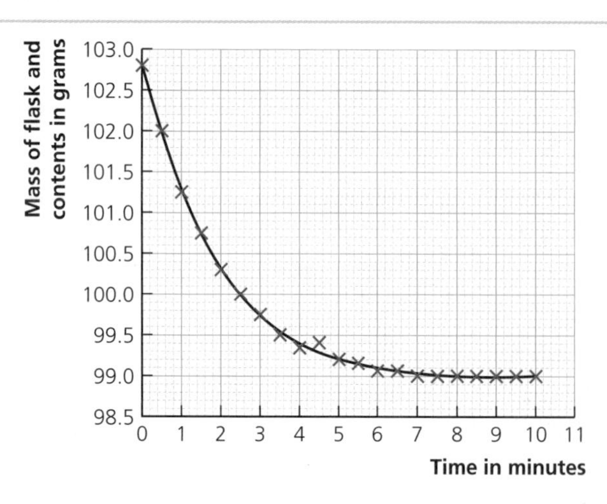

a. Are there any results that you would ignore when drawing a best fit curve?
b. At time 2 minutes what is the mass of the flask and contents?
c. Describe the trend in the graph.
d. How can the effect of random error be reduced in this practical?
e. How would the effects of systematic error be reduced in this practical?

4. Decide whether each of the following will lead to a systematic or random error.
   a. Using an analogue ammeter with a zero error.
   b. Forgetting to re-zero the balance in one repeat experiment.
   c. Finding the density of saltwater when asked to find the density of ethanol.
5. Explain the difference between an *error* and a *mistake* when doing an experiment.
6. What type of error does the technique of *repeat and average* reduce and why?
7. A wire is 98.4 cm long. A student measures the length with four different metre sticks. The results are: 96.8 cm, 97.1 cm, 97.7 cm, 97.9 cm
   a. What type of error has been made by the student?
      Give a reason for your answer.
   b. Three other students measure the same length of wire with different metre sticks and they all get 98.4 cm.
      Which two of the following words apply to these measurements?
      reproducible, repeatable, valid
8. Why are reproducible results less likely to contain systematic errors than results that are just repeatable?
9. What do large range bars on a graph indicate?

Scientific vocabulary, quantities, units and symbols

Science vocabulary

Science has its own vocabulary. Words used often have a very specific meaning. So, it is important to learn the definitions of technical terms to understand and create your own written science. A common mistake is to use words as they are understood by the non-scientist. For example, stating that the weight of an object is 50 kg, when you should be talking about mass. For information on scientific vocabulary used in GCSE Science, see Tables 2.2, 2.3 and 2.4 on page 87.

Science quantities, units and symbols

For information on the quantities and units used in science, as well as how to convert between them, refer to the Maths skills chapter of this book.

4 Revision skills

This chapter covers the importance of revision and the key strategies that can help you gain the most benefit from your revision. A common misconception is that there is only one way to revise – one that involves lots of note-taking, re-reading and highlighting. However, research shows that this is not an effective way of revising. You need to vary the techniques you use – and find the ones that work best for *you* – to make the most of revision.

Students often think they can change the way they revise, or that revision is something you either can or cannot do. In fact, revision is an important skill and, like any skill, with support and practice you can get better at it. By 'better', this means that you can revise more efficiently (in other words, you'll get a greater benefit from the same amount of revision time) and more effectively (in other words, you'll retain more information).

This chapter will cover the key elements of successful revision:

- Planning ahead
- Using the right tools
- Creating the right environment
- Useful revision techniques
- Practice, practice, practice!

» Planning ahead

The key to successful revision is planning. There are a number of things to bear in mind when planning revision.

Be realistic

There is nothing more demotivating than setting unrealistic targets and then not fulfilling them. You need to think carefully about how much work you can realistically complete and set a reasonable time to complete it.

Ensure you cover all topics in the course

It is tempting to focus on what you think are the most important areas and leave out others. This is risky because no one knows what you will be asked about. It's a horrible feeling seeing an exam question on a topic that you know you haven't revised. In this section, there is advice on the sorts of strategies to ensure you cover all the key points of the specification.

Tip

Spend a small amount of time each evening during your GCSE course going over what you learnt in that day's lesson – it can be really beneficial. It helps you remember the content when you come to revise it, and provides good preparation for the next lesson.

Make friends with the areas you don't like

It is tempting to focus on the areas you already know and are good at. It makes you feel like you're making great progress when, in fact, you're doing yourself a disservice. You should work hard at the areas you find difficult to make sure you give yourself the best chance. This can be tough as you may feel progress is slow, but you must persevere with it.

Ask for help

The most successful students are often those who ask questions from teachers, parents and other students. If there is anything on the specification that you are unsure about, don't stay silent – ask a question! Proper planning will ensure you have time to ask these questions as you work through your revision.

Target setting

Targets are an important part of successful revision planning. You may want to include SMART targets in your revision timetable.

Here's an example of a SMART (specific, measurable, achievable, realistic and timely) target.

Target: Achieve at least a grade 6 in a practice Chemistry Paper 1 done under exam conditions. This should be completed by the end of the week.

- **Specific** – this target is specific as it gives the exam paper, how it needs to be completed and the grade required.
- **Measurable** – as a specific minimum grade is given (6), this target is measurable.
- **Achievable** – as long as there is time to complete the paper, which there should be if it's being completed in the 'time allowed', then this target would be achievable.
- **Realistic** – you shouldn't be expecting to score grade 9 in assessments straight away or learn huge amounts of content in a very small time; so, a grade 6 seems to be realistic for a first stab.
- **Timely** – there is a set time to complete this goal, namely by the end of the week. Assuming that the student has revised all of the topics on this paper by then, this is a sensible timeframe.

Targets can also be smaller and set for individual revision sessions, for example:

- complete three practice questions on one maths skill
- get 75% on a recall test
- learn the stages of a process, e.g. the carbon cycle
- make a set of key word flash cards on Lenses and Visible Light

Setting targets for each revision session will help you realise when you are finished, as well as providing yourself with evidence of your progress – always a good motivator!

Tip

Targets can include things such as not using social media or your phone for a whole revision session if this is something you particularly struggle with.

» Using the right tools

Having the right tools is vital for effective revision. Some of the 'practical' tools you'll need during your revision would include:

- a planner or diary
- pens
- paper
- highlighters
- flash cards
- and so on ...

Having these tools close to hand will remove simple barriers to successful revision – such as not having a pen!

Revision timetables

Revision timetables are a useful tool to help you organise and structure your work. Remember that the key is to be realistic – don't plan to do too much, or you'll become demoralised.

Revision works best in shorter blocks. So, don't plan to spend two hours solidly revising one topic – you probably won't last that long. Even if you do, it's unlikely the work towards the end of this time will be effective.

If you are making a revision timetable for mock exams (before you've finished your course), you will need to allow time for any homework set in addition to revision.

How to create a revision timetable

Identify the long-term goal and short-term targets you're trying to achieve (and make sure they're SMART). Ask yourself if this a general timetable to use during the term, or one aimed at preparing for a particular exam or assessment. This will affect how you build your plan as your commitments will vary.

Whatever the end goal, don't plan so you only just finish in time. Make sure you plan to cover all the topic areas you need well before the assessment. That way, if you encounter problems that slow you down, you won't run out of time.

Tip

Include your other commitments in a revision timetable, such as music lessons, sports, exercise or part-time work. This will give a clearer picture of how much time you have for revision. These commitments could be rewards – they give you something look forward to. Or it may become clear that you may have too much on and have to (temporarily) give something up.

Tip

Make sure you carefully plan how much time you have available before each exam. Miscalculating by even a week could cause problems.

Examples of revision timetables

Good example

Revision sessions split into small sections. This helps maintain engagement during the session.

Regular breaks scheduled and realistic expectations of how much revision can be completed in a day.

Times	Mon
8:30am–3:20pm	School
4:00pm–4:30pm	Chemistry (size and mass of atoms)
4:30pm–5:30pm	Football
5:30pm–6:00pm	Dinner
6:00pm–6:30pm	Physics (black body radiation)
6:30pm–7:00pm	Online gaming
7:00pm–7:30pm	Biology (meiosis)

Specific topics given for revision sections – while you don't need to necessarily rigidly stick to this it's good to have a topic focus for each revision session, you can then set targets for the session based around this particular topic area.

Bad example

Unrealistic expectations – timetabling so revision starts at 6:30 am and finishes at 11:00 pm at night is unrealistic and potentially harmful. Failing to achieve set goals can be very demotivating.

Working excessive long hours without adequate sleep and relaxation time can be detrimental to health.

Times	Mon
6:30am–7:20am	Physics
8:30am–3:30pm	School
3:30pm–5:00pm	Physics
5:00pm–7:30pm	Chemistry
7:30pm–11:00pm	Biology

No breaks scheduled – planning breaks, both as a rest and reward, are very important for effective revision.

No specific topics mentioned – 'Physics' is far too vague; what areas are they specifically going to work on?

Long blocks of one subject – the student is unlikely to remain engaged for this length of time.

Revision checklist

A revision checklist is an important tool to ensure you are covering all the required specification content. Your teacher may provide you with a revision checklist, but even if they do, making one yourself can be a useful learning activity.

> **Tip**
> Some revision guides (like *My Revision Notes*) also have checklists included that you can use.

How to make a revision checklist

1. Read the specification; this is everything you need to know.
2. Split the specification into short statements and place them into a grid.
3. Work through the grid, ticking as you complete each stage for a particular topic. Use practice exam questions to check that your revision has been effective.
4. Return to the areas you are weaker in and focus on improving them.

Example revision checklist

The following is an example statement taken from a GCSE Physics specification. This statement has been used as the basis for an example revision checklist.

Learners should have a knowledge and awareness of the advantages and disadvantages of renewable energy technologies (e.g. hydroelectric, wind power, wave power, tidal power, waste, solar, wood) for generating electricity. Learners should also be able to explain the advantages and disadvantages of non-renewable energy technologies, including fossil fuels and nuclear for generating electricity.

Revision checklist

Specification statement	Covered in class	Revised	Completed example questions	Questions to ask teacher
Advantages and disadvantages of renewable energy resources for generating electricity 1 – hydroelectric, wind power, wave power, tidal power.				
Advantages and disadvantages of renewable energy resources for generating electricity 2 – waste, solar, wood.				
Advantages and disadvantages of fossil fuels for generating electricity.				
Advantages and disadvantages of nuclear power for generating electricity.				

Posters

You could create posters of key processes, diagrams and points and put them up around the house so you can revise throughout the day. Be sure to change the posters regularly so that you don't become too used to them and they lose their impact. See the next section for more on making the most of your learning environment.

> **Tip**
> Don't procrastinate by focusing too much on the appearance of your notes. It can be tempting to spend large amounts of time making revision timetables and notes that look nice, but this is a distraction from the real work of revising.

Technology

There are many ways to use technology to help you revise. For example, you can make slideshows of key points, watch short videos or listen to podcasts. The advantage of creating a resource yourself is that it forces you to think about a particular topic in detail. This will help you to remember key points and improve your understanding. The finished products should be kept safe so you can revisit them closer to the exam. You could lend your products to friends and borrow ones they've made to share the workload.

Making your own video and audio

If you record yourself explaining a particular concept or idea, either as a video or podcast, you can listen to it whenever you want. For example, while travelling to or from school. But make sure your explanation is correct, or you may reinforce incorrect information.

Revision slideshows

Slideshows can incorporate diagrams, videos and animations from the internet to aid your understanding of complex processes. They can be converted into video files, printed out as posters, or viewed on screen. It's important to focus on the content of the slideshow – don't spend too long making it look nice.

Social media

Social media contain a wide range of revision resources. However, it is important to make sure resources are correct. If it's user-generated content, there's no guarantee the information will be accurate.

Study vloggers and other students on social media can provide valuable support and a sense of being part of a wider community going through the same pressures as you. However, don't compare yourself to other people in case it makes you feel as if you're not keeping up.

> **Tip**
>
> Some students find listening to music helpful when they're revising, even associating certain artists or songs with specific topics. However, music can also be distracting, so only use it if it works for you.

> **Tip**
>
> Be aware that social media can also be distracting. It's easy to procrastinate if you're not focused. Advice on reducing distractions can found on page 126.

» Creating the right environment

The importance of having a suitable environment to revise in cannot be underestimated – you can have the best plan and intentions in the world, but if you're watching TV at the same time, or you can't find the book you need, or you're gasping for a drink and so on, then you're likely to lose concentration sooner rather than later. Make sure you create a sensible working space.

Work area and organisation

It is difficult to concentrate with the distraction of an untidy work area – so keep it tidy! It is also inefficient, as you may spend time looking for things you've mislaid.

The importance of organisation extends to your exercise books and revision folders. You will have at least two years' worth of work to revise and study. Misplacing work can have a negative effect on your revision.

Put together a revision folder with all your notes, practice questions, checklists, timetables and so on. You could organise it by topic so it's easy to find particular information and see the work you have already completed.

Looking after yourself

Revising for exams is a marathon, not a sprint – you don't want to burn out before you reach the exams. Make sure you stay healthy and happy while revising. This is important for your own wellbeing, and helps you revise effectively.

Eat properly

Try to eat a healthy, balanced diet. Keep some healthy snacks nearby so that hunger doesn't distract you when revising. Food high in sugar is not ideal for maintaining concentration, so make sure you're sensible when selecting snacks.

Drink plenty of water

Make sure you have enough water at hand to last your revision session. It's vital to stay hydrated and getting up to get a drink can be a distraction, particularly if you wander past the TV on the way.

Consider when you work most effectively

Different people work better at different times of the day (morning, afternoon, early evening). Try to plan your revision during the times you're most productive. This may take some trial and error at the start of your revision.

Make sure you get enough sleep

Lack of sleep can lead to serious health problems. Late-night cramming is not an effective revision technique.

Tip

Even if you're more productive in the evening, you still need to go to bed early enough to get enough sleep.

Avoiding distractions

Social media and other technology can provide an unwelcome temptation when studying. Possible solutions to this distraction include:

Plan specific online activities during study breaks

This could be social media time, videos or gaming. This can also give you something to look forward to while you're working. Be careful to ensure you stick to the allotted break time and don't fall into the trap of 'just one more' video or game.

Switch technology off

Switching the internet off can be the most powerful productivity tool. Turn off your phone and consider avoiding the internet whilst studying, only turning them back on at the end of the study session or during a break. This removes the temptation to constantly check your phone or messages. If you do need access to a device while studying, there are a number of blocker apps and services that can limit what you are able to access.

Tell your family and friends

Make sure people know that you are planning to study for a specific period of time. They'll understand why you may not be replying to their messages and they will help you by staying out of your way. This can also help with positive reinforcement, as you can talk to them afterwards about the successful outcomes of the revision session.

» Useful revision techniques

Many students start their GCSE studies with little idea of how to revise effectively. There are many effective revision techniques that are worth trying. And remember, revision is a skill that needs to be learnt and then practised. It may take time to get to grips with some of these strategies, but it will be worth it if you put the effort in.

Memory aids

Before we get onto the revision techniques themselves, here are some tips on how to memorise particularly complicated information. Look out for opportunities to put these techniques into action.

Elaboration

Elaboration is where you ask new questions about what you have already learnt. In doing this you will begin to link ideas together and develop your holistic understanding of the subject. The more connections between topics your brain makes automatically, the easier you will find recalling the relevant information in the exam.

For example, if you have just consolidated your notes on the structure of the plant transport system you might challenge yourself to make a list of all the similarities and differences between the transport systems of plants and humans.

This is useful because, by answering this type of question, your brain will form links between the topics and strengthen your recall while also improving your understanding of both plant and human transport systems.

As part of elaboration you can try and link ideas to real world examples. These will develop your understanding and help you memorise key facts. For example, when revising polymer structure in Chemistry, you could relate this to examples of polymers and how they're used.

Key term

Holistic: When all parts of a subject are interconnected and best understood with reference to the subject as a whole.

Tip

It is helpful to use these types of questions to create linked mind maps showing the connections between topic areas.

Tip

When it comes to mnemonics, the sillier the phrase, the better – they tend to stick in your head better than everyday phrases.

Mnemonics

Mnemonics are memory aids that use patterns of words or ideas to help you memorise facts or information. The most common type is where you create a phrase using words whose first letters match the key word or idea or you are trying to learn.

For example, living things can be classified into these taxonomic levels:

- **K**ingdom
- **P**hylum
- **C**lass
- **O**rder
- **F**amily
- **G**enus
- **S**pecies.

A mnemonic to help remember the order of these levels might be:

King **P**hilip **C**ame **O**ver **F**or **G**reat **S**amosas

Other mnemonics include rhymes, short songs and unusual visual layouts of the information you're trying to remember.

Memory palace

Memory palace is a technique that memory specialists often use to remember huge amounts of information. In this technique, you imagine a place (this could be a palace, as in the name of the technique, but it could be your home or somewhere else you're familiar with), and in this location you place certain facts in certain rooms or areas. These facts should, ideally, be associated with wherever you place them, and always stay in the same location and appear in the same order.

You may also find it helpful to 'dress' each fact up in a visual way. For example, you might imagine the information 'gravitational acceleration is ~10 m/s^2' being 'dressed' as the apple that fell on Newton's head. You might then place this apple in the kitchen in your memory palace, 10 metres high on top of one of your cupboards.

Through the process of associating facts and their imaginary location, you are more likely to correctly recall the fact when you come to revisit the 'palace' and locations in your mind.

Make your revision active

In order to revise effectively, you have to actually *do something* with the information. In other words, the key to effective revision is to make it active. In contrast, simply re-reading your notes is passive and is fairly ineffective in helping people retain knowledge. You need to be actively thinking about the information you are revising. This increases the chance of you remembering it and also allows you to see links between different topic areas. Developing this kind of deep, holistic understanding of the course is key to getting top marks.

Different active techniques work for different people. Try a range of activities and see which one(s) work for you. Try not to stick to one activity when you revise; using a range of activities will help maintain your interest.

Key term

Active revision: Revision where you organise and use the material you are revising. This is in contrast to passive revision, which involves activities such as reading or copying notes where you are not engaging in active thought.

» Retrieval practice

Retrieval practice usually involves the following steps.

Step 1 Consolidate your notes

Step 2 Test yourself

Step 3 Check your answers

Step 4 Repeat

Step 1: Consolidate your notes

Consolidating notes means taking information from your notes and presenting it in a different form. This can be as simple as just writing out the key points of a particular topic as bullet points on a separate piece of paper. However, more effective consolidation techniques involve taking this information and turning it into a table or diagram, or perhaps being more creative and turning them into mind maps or flash cards.

Bullet point notes

Here is an example of how you might consolidate bullet point notes from a chunk of existing text.

Original text

Ultrasound waves are inaudible to humans because of their very high frequency. These waves are partially reflected at a boundary between two different media. The time taken for the reflections to echo back to a detector can be used to determine how far away this boundary is, provided we know the speed of the waves in that medium. This allows ultrasound waves to be used for both medical and industrial imaging.

Seismic waves are produced by earthquakes. Seismic P-waves are longitudinal and they travel at different speeds through solids and liquids. Seismic S-waves are transverse, so they cannot travel through a liquid. P-waves and S-waves provide evidence for the structure and size of the Earth's core. The study of seismic waves provides evidence about parts of the Earth far below the surface.

Consolidated notes

- Ultrasound frequency > 20 000, so humans can't hear them
- Ultrasound reflects and the time it takes an echo to return can be used to find distance between target and source
- Ultrasound is used in medicine and industry to obtain images

- Two types of seismic wave in earthquakes: longitudinal P-waves and transverse S-waves
- P-waves can travel through solids and liquids, S-waves through solids only
- Both give information about interior structure of the Earth, e.g. size of the core

Flow diagrams

Flow diagrams are a great way to represent the steps of a process. They help you remember the steps in the right order. An example of a Chemistry flow diagram, for the Haber process, is shown in Figure 4.1.

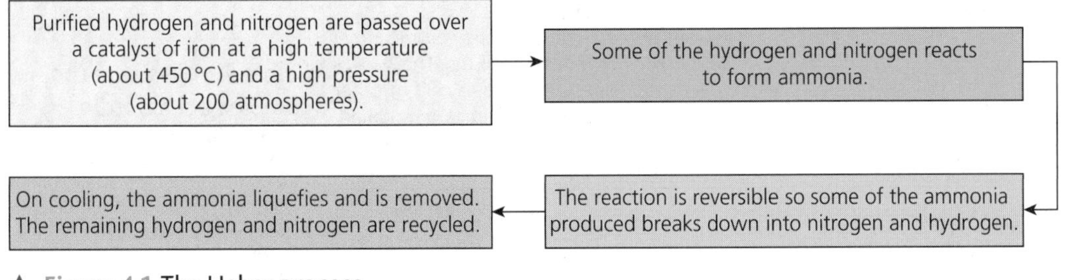

▲ Figure 4.1 The Haber process

Mind maps

Mind maps are summaries that show links between topics. Developing these links is a high-order skill – it is key to developing a full and deep understanding of the specification content.

Mind maps sometimes lack detail, so are most useful to make once you have studied the topics in greater detail.

See **Elaboration** (page 127) for more information on the importance of linking ideas in active revision.

> **Key term**
> High-order skill: A challenging skill that is difficult to master but has wide ranging benefits across subjects.

Good example of a mind map

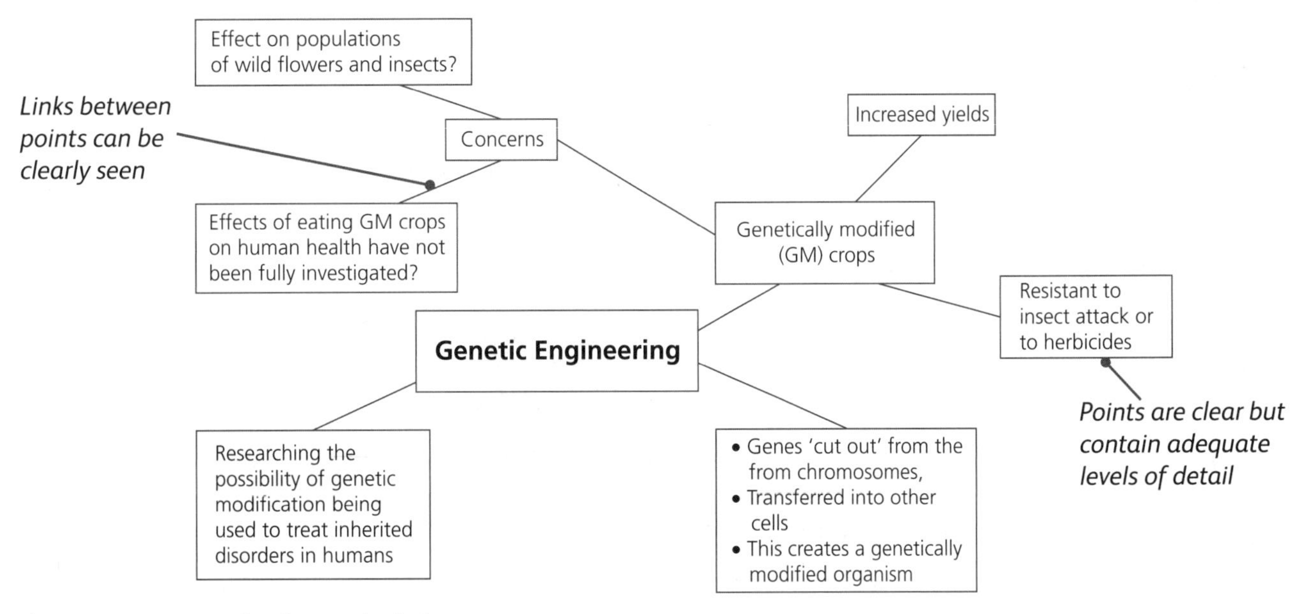

▲ Figure 4.2 Example of a good mind map

Bad example of a mind map

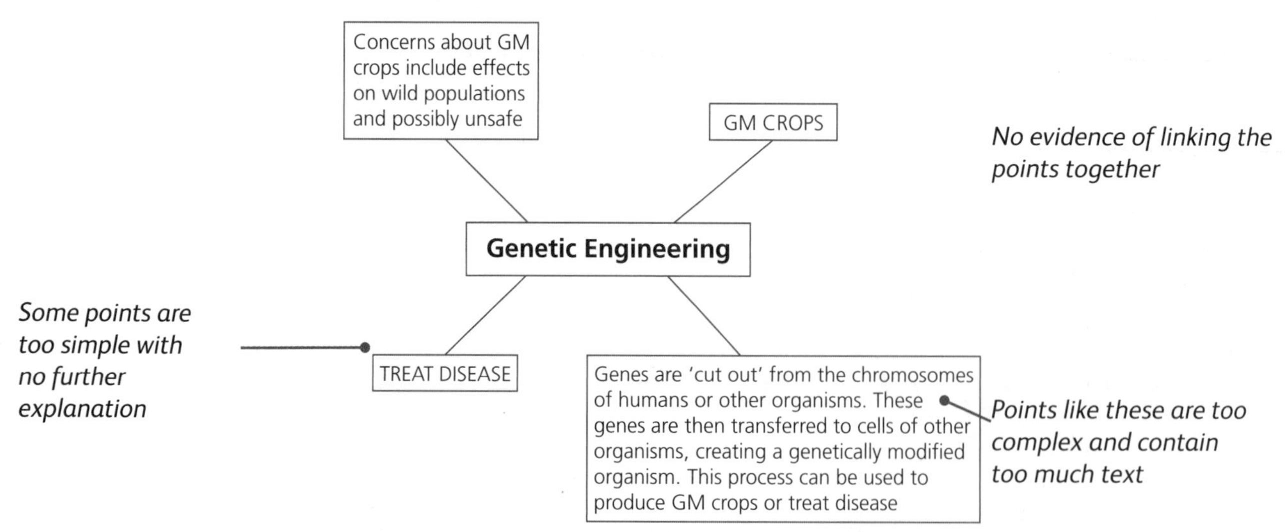

▲ Figure 4.3 Example of a bad mind map

Flash cards

Flash cards are excellent for things such as key word definitions – write a key word on one side of the card and the definition on the other.

Flash cards can also be used to summarise key points of a process or topic area.

Similar to mind maps, they should be used in conjunction with other revision methods that fully cover the detail required.

> **Tip**
> Like with mind maps, do not squeeze too much information on a flash card.

Step 2: Test yourself

There is a range of test activities you can do with the notes you've consolidated, including:

- making your own quizzes
- asking friends or family to test you
- picking flash cards randomly from a stack
- trying past exam questions

Whether you've created a test, or are asking other people to test you, it's important that you leave a decent period of time between consolidating your notes and being tested on it. Otherwise, you are not effectively testing your recall.

> **Tip**
> There are a number of different apps that are useful to help with creating quizzes. Some of these apps also allow you to share these quizzes with friends, so you can help each other out.

Step 3: Check your answers

After testing yourself, check your answers using your notes or textbooks. Be hard on yourself when marking answers. An answer that's *almost* right might not gain full credit in an exam. You should always strive to give the best possible answer.

If you get anything wrong, correct your answers on paper (not just in your head). And annotate your answers with anything you've missed along with additional things you could do to improve, such as using more technical language.

Step 4: Repeat

Repeat the whole process, for each topic, at regular intervals. Revisiting activities will help you memorise key aspects and ensure you learn from your previous mistakes. It is especially helpful for topics you find challenging.

> **Tip**
> Even though spacing out and mixing up topics are separate sub-sections here, they should be incorporated into your retrieval practice.

When repeating, do not *immediately* revisit the same topic again. Effective revision is more likely if you leave time before revisiting topics you've recently revised, and use this time to mix in other topics.

Spacing out topics

Once you've gone through a whole topic, move on and wait before returning to it and testing your recall. Ideally, you should return to a topic regularly, increasingly long intervals between each return. Returning to a topic needn't take too long – quickly redoing some tests you took before may be enough.

When you return, ask yourself:

- do you know the topic as well as you did when you revised it first time around?
- are you still making the same mistakes?
- what can you improve on?

Identify the key areas you need to go back over.

Allow time for this revisiting process in your revision timetable. Leaving things to the last minute and trying to cram is not an effective way of revising.

> **Tip**
> This is sometimes referred to as 'Spaced Practice'.

Mixing up topics

Mixing up topics (covering a mix of topics during your revision timetable rather than spending long periods of time on one) is an effective revision strategy. It ties into the need to revisit topics at intervals. Mixing up and revising different areas means it's inevitable there will be a space between first revising a topic and then coming back to it a later date.

Studies have shown that, although moving onto different topics more regularly may seem difficult, it could significantly improve your revision. So it's worth persevering.

> **Tip**
> This is sometimes referred to as 'Interleaving'.

» Practise, practise, practise

Completing practice questions, particularly exam-style questions, allows you to apply your knowledge and check that your revision is working. If you're spending lots of time revising but finding you cannot answer the exam questions, then something's wrong with your revision technique and you should try a different one. Examples of practice questions can be found on pages 150–160.

Practice exam questions can be approached in a number of ways.

Complete the questions using notes

This may seem a bit like cheating but it is good, active revision and will show you if there are any areas of your notes that need improving.

Complete questions on a particular topic

After revising a topic area, complete past exam questions on that topic without using your notes. If you find you get questions wrong, go back over your notes before returning to complete questions on this topic area again at a later date. Repeat this process until you are consistently answering all the questions correctly. Annotate your revision notes with points from the mark schemes. More details on the use of mark schemes can be found on pages 90–101.

Complete questions on a topic you have not yet revised fully

This will show you which areas of the topic you know already and which areas you need to work on. You can then revise the topic and go back and complete the question again to check that you have successfully plugged the gaps in your knowledge.

Complete questions under exam conditions

Towards the end of your revision, when you're comfortable with the topics, complete a range of questions under timed, exam conditions. This means in silence, with no distractions and without using any notes or textbooks.

It is important to complete at least some timed activities under exam conditions. The point of this is to prepare you for the exam. Remember, if you spend time looking up answers, talking, looking at your phone and so on, you won't get an accurate idea of timings.

Always ensure you leave enough time to check back over all your answers. Students often lose lots of marks due to silly mistakes, particularly in calculations. These can be avoided by ensuring you check all answers thoroughly.

When working your way up to completing an exam under timed conditions, it can be helpful to begin with timing one or two questions to get yourself used to the speed at which you should be answering them. You can then slowly work your way up to completing full-length papers in the time you would have in the real exam. Make a note of the areas where you found you were spending too long and look for ways to improve.

Effective revision is absolutely vital to success in GCSE Science. As you are studying a linear course you'll be examined on a whole two years' worth of learning. Only by revising effectively and thoroughly can you ensure you have a full and complete understanding of all the content.

Tips

As a guide to timings, you can work out how many marks you should be ideally gaining per minute. To do this, divide the total number of marks available by the time you have in the exam. This will help you get an idea of what questions need longer, but it is not a perfect guide as some questions will take longer than others, particularly the more complex questions that are often found towards the end of the exam paper.

5 Exam skills

You will have spent at least two years learning GCSE Science by the time you take the exam. Given all the work that you have put into your studies, it is important that you know how to apply your knowledge in exam conditions.

Learning the content in the specification is only part of being successful. You must also develop your exam skills to ensure that you get the highest marks you can. This includes being fully prepared before the exam, being aware of the types of questions and the command words used so you know what you need to do, and simple things like checking your answers.

This chapter shows you how to prepare for the exam, how to understand what each question is asking and how to judge the level of content you need to write to get maximum marks.

» General exam advice

Before the exam

Exam specifics

To make sure there are no nasty surprises in the exam, you should read up on how your particular exam board will examine you. If you don't know your exam board or how to access its website or specification, ask your teacher. You also should be aware that most exam boards have different GCSE specifications, and you need to find out which qualification you are doing, for example whether you are doing a Double or Single Award. Some boards also do completely separate qualifications (for example OCR offers both Gateway Science and Twenty-First Century Science).

The types of things you need to look out for are:

- how many papers you will be sitting
- how the papers are split up (in terms of marks and content)
- how long each paper lasts
- whether there are any other assessments (for example a minority of exam boards assess practical work independently).

The subject content section of the specification is also useful for telling you everything you must know, understand and be able to do. Many revision guides, such as *My Revision Notes*, will have checklists for what you need to cover, so you can tick them off as you familiarise yourself with each area.

> **Tip**
> Write the number and length of each paper on a post-it and stick it near your desk, to remind you of what you are working towards. A list of topics featured on each paper is also handy to have.

Sample assessment materials

Sample assessment materials and past exam papers are an incredibly useful resource. Past papers will show you the style of questions you can expect to be asked. For each paper, you should also check the mark scheme to see exactly how each question is marked. You will already be familiar with how some questions are marked if you have worked through the Extended Responses chapter of this book (beginning on page 84).

These materials can be accessed through the exam board's website, although the most recent ones may not be publicly accessible as they will be on a secure part of the site. You can ask your teacher if they can download them for you, but they may want to save them to do in class or set as homework.

Tip

Practice exam questions can also be found on pages 150–160 of this guide.

Practise answering full past papers to help you get used to the length of paper and examination style presented by your examination board. Study the mark schemes carefully and ensure that you note which points you did not include in your own answer. Mark schemes show exactly what the examiner is looking for and they will enable you to take a more detailed, precise and focused approach, which will help you improve your examination technique and answering style.

There are other sources of questions too:

- Exam questions from old versions of specifications – these are usually freely available, and there will probably be a large number of them. These can be very useful because they will cover many of the topics and skills that are assessed in the current specifications. However, you need to be careful because some of the content will have changed, and the question styles may be different too.
- Questions from other exam boards – again, these can be useful if you have already completed all of the available questions for the board you are doing. Exam papers for the recent specifications will contain more of the application-style questions that are a feature of most new specifications. As with old specifications, you need to be careful not to rely too much on these resources, and only complete questions that match the content of your specification.

Planning ahead

Exams are stressful, so it is very important to reduce stress on the day as much as possible. There are a few ways you can do this:

- Your school will give you your timetable in the summer term, but you can download the full timetable from the exam board's website long before that, if you wish. This will help you plan ahead and make revision schedules.
- Make sure that you get all the equipment you need ready in plenty of time. It might even be worth packing the night before. This means sorting out your pens, pencils, rulers, calculator, etc. Make sure you have got spares of everything, in case anything runs out or breaks during the exam. You can't bring your mobile into the exam hall – unless your school has special arrangements, it's best to leave it at home.
- Make sure you know where your exam is taking place, and your seat number. You should also bring your statement of entry. This will reduce any chance of turning up at the wrong venue. It will also make it much easier when you arrive at the exam.
- Make sure you know how you are going to get to your exam, and ensure you plan to arrive in plenty of time. Getting stuck in traffic is only going to increase your anxiety and make it more difficult to perform at your best.
- Make sure you get enough sleep the night before the exam. This will improve your concentration in your exam. Late-night cramming the night before is rarely effective (see Section 4 of the book for further revision tips).

During the exam

Understanding what to do

Once you're in the exam room, and have been given the question paper, you should read the advice and instructions on the front cover. Complete the candidate details when told to do so.

As you work through the paper, read each question carefully. Look at command words that tell you what you have to do.

If it is a long question, plan your answer carefully before you start to write.

If it is a mathematical question, think formula, substitutions, calculation, answer with unit.

Look carefully at the space provided for your answer. The amount of space will give you a hint on the maximum you are expected to write to get full marks.

Tip

Although the amount of lines is a useful guide, do not feel like you have to fill all the space if you are confident that you have written enough for the marks in a shorter space.

Time management

Time management in exams is vitally important. As part of your preparations, you should have already practised completing exam papers in the time given, and you should know that the time allocation for your GCSE papers gives you about 1 minute per mark – try to stick to these timings. Below is some further advice on time management in the exam:

- Keep an eye on the time – do not get too obsessed with the clock but make sure you are checking regularly to see if you are approximately sticking to your timings. You need to be a bit flexible as you may find that some questions take longer and some take less time, but if you are starting to fall behind, try to speed up.
- Some questions will require more time than others – especially any calculations. If you are struggling with a calculation or a part of a question, leave it out and move on. Then, if you have time at the end you can come back and complete it.
- Find time to answer every question – you should have enough time to answer all questions on the paper, but you need to ensure that you do not include irrelevant detail that may otherwise waste time. Remember that the key to success is obtaining maximum marks in minimum words.
- Your question papers will start with a lower demand question and then will slowly go up in difficulty. Similarly, within a multipart question there will be an easier lead in, building through successive parts of the question. This means you have a fair chance of gaining some marks on each topic area throughout the paper.
- Do not spend too long on a tricky question – if you get to a question that you are really struggling to answer, don't give up, just move on to the next one; it may be easier. Draw a star beside any skipped question and come back to it at the end if there is time. Spending a long time on one harder question can use up valuable time that could be spent on questions you can more easily answer.
- The only certain way of scoring zero marks on a question is by writing nothing. You should try to write an answer for every question even if you do not know where to start. Try noting down key words that apply to the topic in case it either jogs your memory or picks up a mark or two.

Tip

A structured question gets progressively more difficult as you work through it. If you get stuck on a harder part, move on to the next question; it will start with an easy part on a different topic, and build your confidence again.

- Leave time to check your answers – this is very important, even if it is not a very enjoyable task. Basic marks are often lost through obvious mistakes such as missing out a key word, putting the wrong letter down or completing part of a calculation incorrectly. By spotting these mistakes and correcting them, you can gain marks that might make all the difference.

Showing your working

Frequently, examiners complain that students do not show their working. This is probably because you often do the whole of a calculation on a calculator and do not always think to write the steps you used to get to the answer. It is incredibly important to show all working because you can still be awarded some method marks even if you get the final answer wrong.

> **Tip**
>
> You will see in sample mark schemes how examiners give marks for errors carried forward (or ECF). This is a useful indicator of how you can pick up marks even when you make a mistake.

Checking your answers

As well as finding time to do a general check of the accuracy of your answers, you should do a quick check to ensure that you have spelt all words correctly. Generally, if a word is misspelt phonetically it will still be credited, but you should take care over key words and technical terms just in case. This is particularly true for extended response questions, where up to six marks are available.

Go through a quick checklist in your head. Ask yourself:

- Does each sentence begin with a capital letter and end with a full stop?
- Is the answer set out in paragraphs?
- Is the spelling correct (particularly the technical terms)?
- Are technical terms used correctly?

Spelling does particularly matter where there are two or more words that are similar to each other that mean very different things. For example, mitosis and meiosis. Spelling either one of these words wrong could cost you marks.

For this reason, it is also worth making sure that your answers are clear and easy to read. While the examiner will not penalise you for messy handwriting, it is very important that they know what you have written.

In mathematical questions, remember to check your arithmetic and that your answer is reasonable. If, for example, you are asked 'Calculate the mass of the student…', you should know that an answer of 500 kg (half a tonne) or 5 kg (a bag of potatoes) are both unlikely. If you have an answer like this it is probable that you have incorrectly multiplied or divided by 10 somewhere in the calculation.

Other common troublemakers

Here are some common troublemakers that you should look out for in your exams:

Biology

- Not answering the question – working out what to write can be one of the more challenging aspects of answering exam questions. Make sure you read every question in full so that you understand what the command word is asking. You will often be given useful information and guidance in the question, which should also be included.

- Writing too little – sometimes a word or a sentence is enough to answer a question and get the marks. However, if a question is worth two or more marks, you probably need to write a bit more.
- Writing too much – you may want to write down everything that you can think of about a topic, but if it is not relevant, you are just wasting time. You may even make your answer worse because the more you write, the more likely you are to say something incorrect or contradict yourself, which could lose you marks.
- Not using key words – key words, or technical terms, are of vital importance in Biology. Mark schemes will often include key words that have to be in an answer in order to score a mark.

Tips

- Remember that some questions may have more than one command word.
- The number of lines given underneath the question is usually a good guide to how much you are expected to write. This is not perfect as people have different sizes of handwriting, but it is good as an estimate. The number of marks is also key – ask yourself if you have definitely included at least as many points as there are marks available.
- For maths questions involving a calculation, it is very difficult to write too much, therefore make sure you write out all of the steps.
- As part of your revision, you should be making lists of key words and memorising them (for example using flash cards). Then you can think back over these lists in the exam and try to recall if there are any you could include.

Chemistry

- Chemical symbols and formulae – remember that symbols have a capital, followed by a lower-case letter. The use of incorrect symbols will be penalised on your examination paper. For example, the use of 'h' for hydrogen, 'CL' for chlorine or 'br' for bromine will all be penalised. Incorrect formulae will also be penalised, for instance, as Na_2CO_3 is the correct formula for sodium carbonate, even a small error such as labelling it $NaCO_3$ will not be credited.
- Equations – always check that your equations include all the correct formulae and are correctly balanced.
- Organic structure – when drawing structural formulae make sure that each atom is bonded correctly. For example, for an alcohol formula, C–HO is incorrect, it must be written as C–OH.

Physics

- Definitions and laws – write down the key ones that you are expected to recall accurately. Many of the equations you require are provided in the examination, but you must be able to recall all the others and be able to use them.
- Required practicals – make sure you can describe all the experiments you carried out during your course, particularly required (or core) practicals. In Physics, these are often chosen as the subject of six mark extended writing questions.

- Technical use of a calculator – make sure you know how to use your calculator effectively and are well-practised. Are you confident, for example, that you can enter numbers in standard index form, find square roots, express numbers to 1 or 2 significant figures and know the meaning of buttons such as S⇔D?

Understanding assessment objectives

As well as the content you need to learn, the exam board specification also sets out the types of questions that can be asked and the percentage of marks for each type. Assessment objectives set out how your skills and knowledge will be tested in the exam. In science, exam questions come under one of three assessment objectives. These objectives are the same in all exam boards, and are shown in Table 5.1.

Table 5.1 Assessment objectives

Assessment objective	Approximate weighting %
AO1: Demonstrate knowledge and understanding of scientific ideas, scientific techniques and procedures.	40
AO2: Apply knowledge and understanding of scientific ideas, scientific enquiry, techniques and procedures.	40
AO3: Analyse information and ideas to interpret and evaluate, make judgements and draw conclusions, develop and improve experimental procedures.	20

AO1 questions

AO1 questions are usually based around factual recall. These questions are usually worth a small number of marks (unless you are being asked to recall a lot of separate facts). A typical AO1 question might be as follows:

Tip

It is very important to read AO1 questions fully. While they may seem very straightforward, there could be additional details in the question. In this worked example, the difference in size is stated in the question, which means that this cannot be given as an answer. This may seem obvious, but it is surprising how many students lose easy marks this way.

A Worked example

Prokaryotic cells are usually much smaller than eukaryotic cells. State one other difference between prokaryotes and eukaryotes. [1]

Model answer

Eukaryotic cells have genetic material enclosed in a nucleus, while the genetic material of prokaryotic cells is not enclosed in a nucleus.

You will see that these questions are literally just asking you to state information, without going into the subject matter in any more detail.

AO2 questions

Recall of facts is not enough to obtain a good grade; many questions require you to apply your knowledge to different and unfamiliar contexts. AO2 questions aim to assess your ability to apply scientific ideas, theories, scientific enquiry, and practical skills and techniques, to explain phenomena and observations in familiar and unfamiliar context. Often these questions are set in novel theoretical and practical contexts. In terms of application of practical knowledge this could be application of a technique or procedure to a novel situation. It could also be application of investigative skills, for example, data analysis. Maths, including graphs and chemical equations, are also assessed under AO2.

A Worked example

Calculate the mass of sodium hydroxide required to make 1000 cm³ of a 0.25 mol/dm³ solution of sodium hydroxide. **[2]**

Model answer

M_r NaOH = 23 + 16 + 1 = 40

0.25 mol/dm³ means there are 0.25 mol in 1000 cm³

Moles of NaOH = 0.25 = $\frac{\text{mass}}{M_r} = \frac{\text{mass}}{40}$

Mass = 40 × 0.25 = 10 g

Most questions that ask you to write balanced symbol equations, or draw bonding diagrams for covalent or ionic bonding, are also AO2 as they require you to apply your knowledge of the topic to a specific chemical.

AO3 questions

AO3 questions are not as common as the other types, but they are often the most challenging questions on the exam, and are therefore usually worth the most marks. These questions may require you to analyse information and use this analysis to interpret, evaluate or draw conclusions. You may also need to develop your own idea or hypothesis. AO3 questions may refer to new examples that you have not seen before, but these questions are all about applying the knowledge you will have covered in your course to a new context. A typical AO3 question might be as follows:

A Worked example

Describe how you might measure the personal power of a student.

In your method, state the measurements you would make, the measuring instruments you would use, and what you would do with your results to calculate the power. **[6]**

Model answer

1 First, find the weight of a student *W* using bathroom scales calibrated in newtons.
2 Measure the vertical height *h* of a staircase with a measuring tape.
3 With a stopwatch, time how long it takes *t* for the student to run from the bottom of the staircase to the top.
4 The power *P* developed by the student is given by $P = \frac{W \times h}{t}$

» Understanding what command words mean

Tip

Questions that ask you to suggest an improvement to an experimental method are a common type of A03 question.

You should have already come across command words in the Extended Writing and Revision sections of this book. Command words are the words and phrases used in exams that tell you how to answer a question. They will also often hint at what assessment objective they are testing. For example, words like 'State' and 'Identify' usually indicate AO1 questions; words like 'Explain' and 'Calculate' usually indicate AO2 questions; and words like 'Justify' and 'Outline' usually indicate AO3 questions.

Each command word is part of a command sentence, such as: 'Explain how sodium chloride conducts electricity.' The command word almost always occurs at the start of the sentence.

You should always underline the command word in the question and focus on it before you start your answer. It is very easy to lose marks by not doing what the question tells you to do.

The following is a guide to the most common command words and what they are asking you to do, with examples of what a model answer to each might look like:

Command word: Calculate

In a 'Calculate' question, you should use numbers given in the question to work out the answer. A numerical answer is expected. You may also be asked to include the correct units with your answer or to write it to a certain number of significant figures. Sometimes you will need to choose the correct equation to use and substitute the correct numbers into the equation to obtain your answer.

A Worked example

Calculate the relative formula mass of magnesium nitrate $Mg(NO_3)_2$. **[2]**

(relative atomic masses: Mg = 24, N = 14, O = 16)

Model answer

$$\text{Relative formula mass} = 1 \times Mg + (2 \times N) + (6 \times O)$$
$$= 1 \times 24 + (2 \times 14) + (6 \times 16) = 148$$

Note that for relative formula mass there are no units.

This is a model answer because it correctly identifies the formula needed, substitutes in the correct values and completes the calculation accurately. Note how the working has been shown to maximise the chances of gaining marks.

Command word: Choose

In a 'Choose' question, a list of alternatives will be given. You need to select the correct one to answer the question. Read the question carefully. Sometimes it will state that each answer may be used once, more than once or not at all.

A Worked example

Some covalent substances, A–D, are shown below.

A: H–C(H)(H)–H B: H–O–H C: H–N(H)–H D: N≡N

Choose the substance, A, B, C or D, that represents

i **methane** **[1]**

ii **a diatomic element.** **[1]**

Model answer

i A

ii D

This is a model answer because it correctly chooses the information needed from the list. No other information is needed.

Command word: Compare

In a 'Compare' question, you need to describe the similarities and/or differences between things. The key to answering 'Compare' questions is to ensure that you include comparative statements, for example, 'Plant cells have a cell wall while animal cells do not.' This is a comparative statement because it mentions two types of cell and their differences, rather than just mentioning the features of one.

A Worked example

Compare the functions of the left and right ventricles of the heart. **[2]**

Model answer

The right ventricle pumps blood to the lungs while the left ventricle pumps blood to the rest of the body.

This is a model answer because it contains a comparative statement in relation to the functions of both the left and right ventricles, the difference between which is clearly stated.

Command word: Complete

In a question that asks you to 'Complete', you need to write your answers in the space provided. This may be on a diagram, within blank spaces in a sentence or in a table.

A Worked example

Complete the diagram below to show the electronic configuration of magnesium. **[1]**

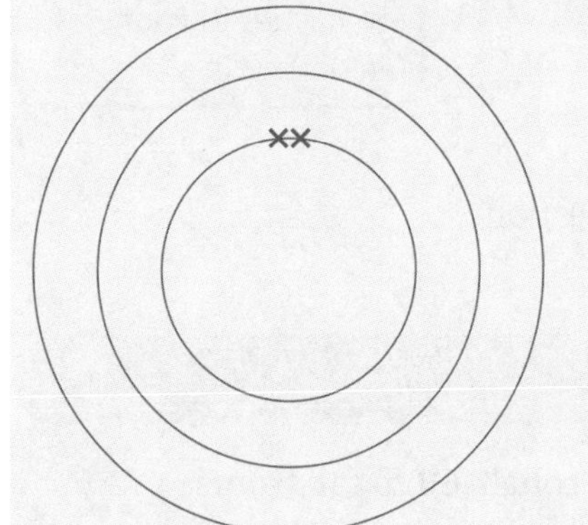

Model answer

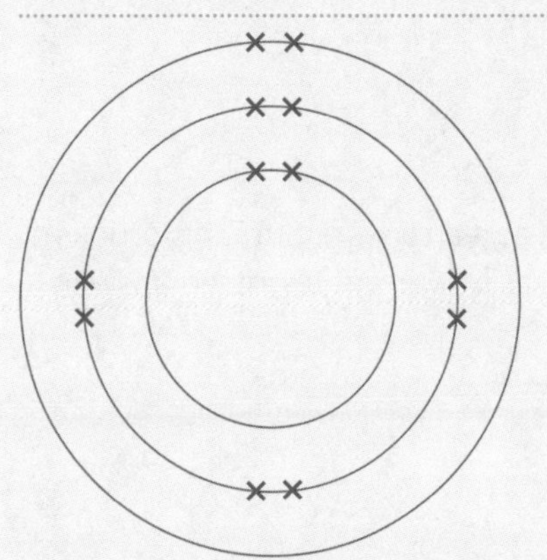

This is a model answer because it completes the diagram that was started, and includes the correct amount of detail and information required by the question.

Command word: Define

Questions that ask you to 'Define' want you to state the scientific meaning of a particular word or phrase.

A Worked example

Define what is meant by the specific heat capacity of a material. **[2]**

Model answer

The specific heat capacity of a substance is the amount of energy required to raise the temperature of one kilogram of the substance by one degree Celsius.

This is a model answer because it provides a simple definition, including correct use and spelling of technical terms.

Command word: Describe

Questions that ask you to 'Describe' want you to recall facts, events or processes, and write about them in an accurate way. For this command word, you just need to describe and not go further. For example, there is no need to explain why something happens. See page 89 for more on how to answer 'Describe' questions.

Tip

Mixing up the terms 'Describe' and 'Explain' is a common mistake – make sure that you read every question carefully. Remember that 'Explain' usually requires you to go further in your answer.

Command word: Design

Questions that ask you to 'Design' want you to set out how something will be done. This will normally be in the context of designing an experiment. See page 92 for more on how to answer 'Design' questions.

Command word: Determine

Questions that ask you to 'Determine' want you to use given data in a question to solve a problem.

A Worked example

Determine how long it would take for the radioactivity of a sample of cobalt-60 to fall from 2560 Bq to 320 Bq if the half-life of the isotope is 5 years. **[4]**

Model answer

$\frac{2560}{320} = 8 = 2^3$; so, three half-lives are required

$3 \times T_{\frac{1}{2}} = 3 \times 5 = 15\,\text{years}$

This is a model answer because it correctly determines the data that you are asked to work out. This problem requires mathematical calculation, but a 'Determine' question might just as easily warrant a written answer.

Command word: Draw

Questions that ask you to 'Draw' want you to produce or add to some kind of illustration. This command word requires you to take a little more time than that required to produce a sketch.

A Worked example

Draw a ray diagram to show how a convex lens can be used to produce an image that is smaller than the object. **[6]**

Model answer

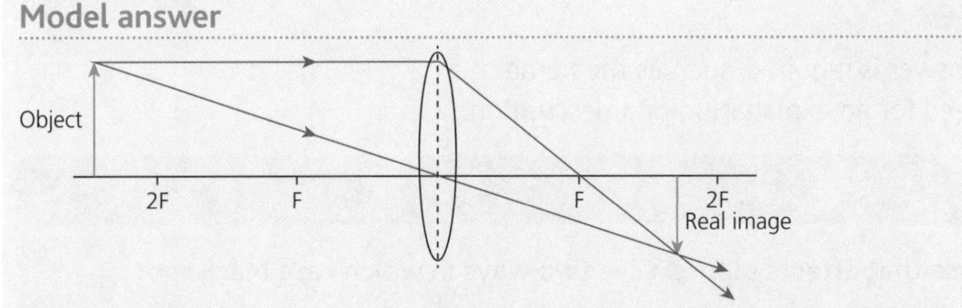

This is a model answer because it is a clear diagram of the image being asked for, with suitable labels and approximate accuracy. If you were asked to 'sketch' an answer instead, you could afford to be a bit more casual, although you should still take as much care in your answer as the time allows.

> **Tip**
> The command word 'Draw' may also be used in questions where you are asked to draw the line of best fit on a graph.

Command word: Estimate

'Estimate' means to give an approximate amount.

A Worked example

The boiling points of some of the halogens are shown below.

Halogen	Boiling point (°C)
fluorine	–188
chlorine	
bromine	60
iodine	184

Estimate the boiling point of chlorine. **[1]**

Model answer

The boiling point of chlorine will be in between that of fluorine and bromine. Choose a number midway, for example –50 °C.

This is a model answer because it provides an answer in the rough ballpark. You don't need to have an exact answer for 'Estimate' questions, although you won't be penalised for using exact figures; but, given the short period of time usually allowed for these types of questions, you are unlikely to have enough time for the full calculation.

Command word: Evaluate

In an 'Evaluate' question, you should use information supplied in the question, and your own knowledge, to consider evidence for and against. This command word will usually be used in longer answer questions, and you should ensure that you give both points for and against the idea you have been asked to evaluate. See page 98 for more on how to answer 'Evaluate' questions.

Command word: Explain

Questions that ask you to 'Explain' want you to make something clear, or state the reasons for something happening. Note the difference between this command word and 'Describe'. 'Explain' is *why* something is happening while 'Describe' is *what* is happening. See page 91 for more on how to answer 'Explain' questions.

Command word: Give

In a 'Give' question, only a short answer is required, such as the name of a process or structure. There is no need for an explanation or a description.

A Worked example

Rose black spot is a fungal disease that affects plants. Give two ways in which rose black spot can be spread. [2]

Model answer

By wind and water

This is a model answer because both methods of the disease spreading are given. This is only a very short answer, but as this is a 'Give' question, no further explanation or description is required.

Command word: Identify

When asked to 'Identify' you should name and/or give the correct formula or symbol.

A Worked example

Identify the substance that is oxidised in the reaction below.

$$Fe_2O_3 + 3CO \rightarrow 2Fe + 3CO_2$$ [1]

Model answer

CO/Carbon monoxide

This is a model answer because, like 'Choose', you are only required to pick the correct answer from the options. Other 'Identify' questions might be more complex than just choosing. For example, questions like 'Identify the independent variable' would require you to recall what an independent variable is and apply it to the information provided. Even so, your answer for any 'Identify' question can be rather short and direct.

Command word: Justify

Questions that ask you to 'Justify' want you to use evidence from the information supplied to support an answer. The key with 'Justify' questions is ensuring that you fully use the information supplied in the question. See page 95 for more on how to answer 'Justify' questions.

Command word: Label

Questions that ask you to 'Label' want you to add text to a diagram, illustration or graph to indicate what particular items are.

A Worked example

The diagram shows a neutral atom. Label the particles identified by the arrows.

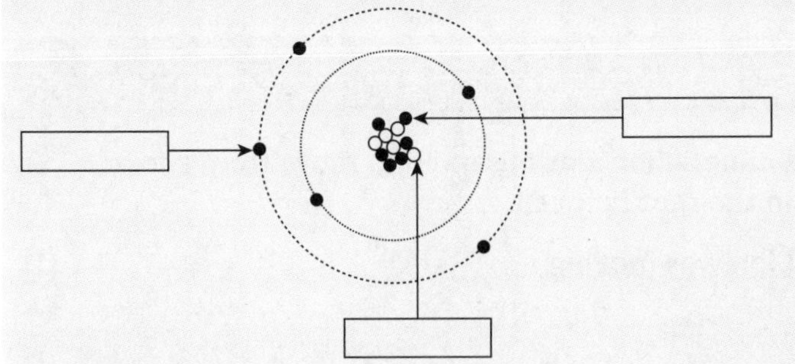

[3]

Model answer

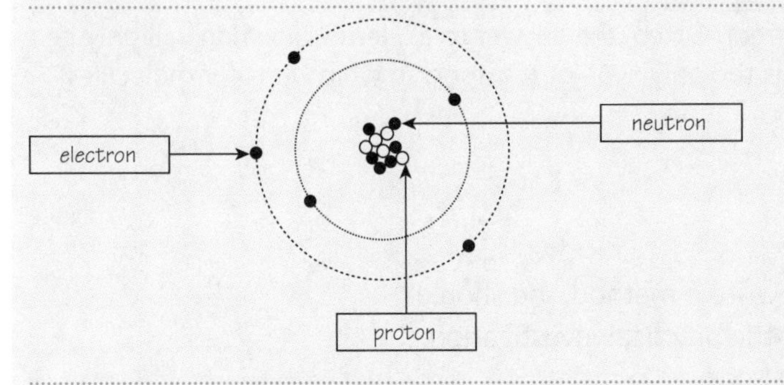

This is a model answer because all the labels are completed clearly. Usually, the labels you have to complete will be signposted as they are in this question, but on rare occasions you may have to draw your own. In this case, make sure it's really clear what the label is pointing at.

Command word: Measure

Questions that ask you to 'Measure' want you to find a figure or data for a given quantity. On rare occasions you may also be asked to use an instrument to determine a particular property.

A Worked example

The diagram shows a ray of light incident on a plane mirror.

Use the protractor to measure the angle of incidence. **[1]**

incident ray

plane mirror

Model answer

Angle of incidence = 64°

This is a model answer because it correctly determines the data that you are asked to work out and includes a unit. In this example an instrument also has to be used correctly to calculate the answer, but this is very rare.

Command word: Name

In questions that ask you to 'Name' something, only a short answer is required – not an explanation or a description.

A Worked example

In an investigation into transport in plant roots, a mineral ion was observed to move from a low concentration in the soil to a high concentration in the root hair cell.

Name the type of transport by which the mineral ion was moving. **[1]**

Model answer

Active transport

This is a model answer because it is concise and correct. Often, the answer to a 'Name' question will only be a single word, phrase or sentence. Active transport is the only type of transport in which ions (or molecules) move from a low to a high concentration.

Command word: Plan

In a 'Plan' question, you will be usually required to write a method. You should write clear and concise points on how to carry out the practical investigation. See page 92 for more on how to answer 'Plan' questions.

Command word: Plot

In a 'Plot' question, you will be required to mark on a graph using data given. Be careful when plotting points or drawing bars as the examiner will check each one. For more information on plotting graphs, see pages 58–60.

A Worked example

An investigation was carried out into the effect of substrate concentration on the rate of reaction of the enzyme lipase. The results are shown in the following table.

Plot the data on a graph. **[4]**

Lipid concentration (%)	Rate of reaction (1/time)
10	0.10
20	0.15
30	0.20
40	0.40
50	0.80

Model answer

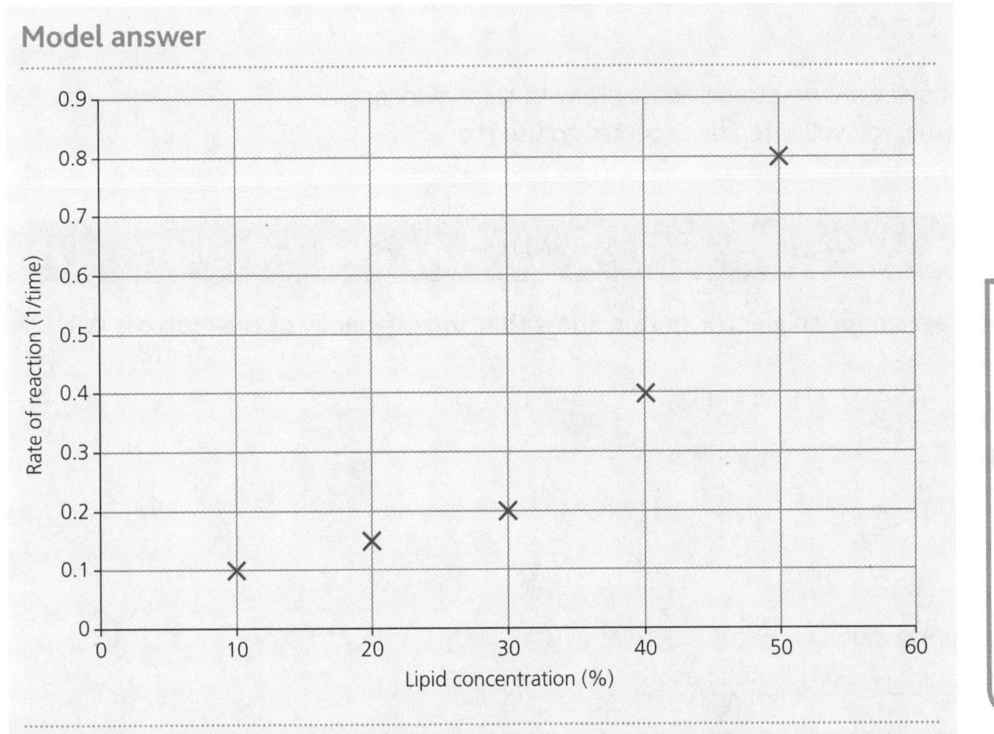

This is a model answer because the two axes are labelled and drawn correctly, and all the graph points are plotted correctly.

Tip

After plotting points on a graph, you may be asked to draw a line of best fit. This can be a straight line or a curved line. You do not have to draw a line of best fit if you have not been asked to.

Command word: Predict

This question expects you to give a plausible outcome. Data is often given in this type of question and you should study it to work out any trends that help with your prediction.

A Worked example

The formulae of three different alkanes are shown in the table.

Name	Formula
ethane	C_2H_6
propane	C_3H_8
butane	C_4H_{10}

Predict the formula of the alkane that contains 5 carbon atoms. **[1]**

Model answer

C_5H_{12}

This is a model answer because it provides a plausible outcome from analysing the information given. You do not have to provide the reasoning for your prediction unless the question asks for it.

Command word: Show

Questions that ask you to 'Show' want you to demonstrate with clear evidence that the statement given is true. You will often be expected to use the information provided.

A Worked example

Below is an energy flow diagram for an electric motor. Show that the efficiency of the motor is 0.8 (80%). **[4]**

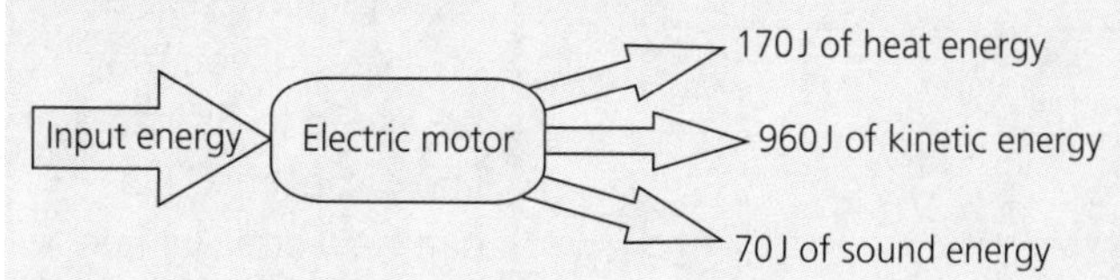

Model answer

total energy input = total energy output = (170 J + 960 J + 70 J) = 1200 J

$$\text{efficiency} = \frac{\text{useful energy output}}{\text{total energy input}} = \frac{960\text{ J [kinetic energy]}}{1200\text{ J}} = 0.8 = 80\%$$

This is a model answer because it quotes the correct equation, identifies the useful energy output and the total energy input, and correctly carries out the calculation. This would, therefore, be considered suitable evidence for 'showing' the statement about efficiency to be true.

Command word: Sketch

Questions that ask you to 'Sketch' want you to produce some kind of illustration, which may be produced quickly.

A Worked example

Sketch a labelled diagram to show the four forces acting on a crate when it is dragged across a rough wooden floor. **[4]**

Model answer

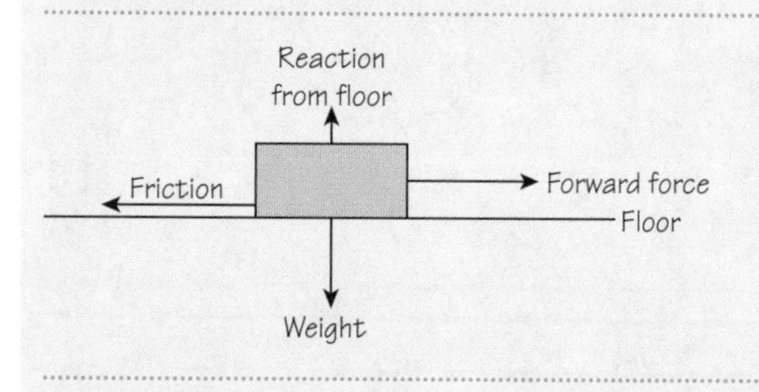

This is a model answer because it shows all the information requested by the question. The fact that it is not overly realistic is not penalised because of the 'Sketch' command word.

Command word: Suggest

In a 'Suggest' question, you will need to apply your knowledge and understanding to a new situation.

A Worked example

An investigation was carried out into the effect of membrane thickness on diffusion of oxygen. Five different membrane thicknesses were used. Each membrane was tested once to determine the rate of diffusion of oxygen.

Suggest a method for making the results of this investigation more reliable. **[3]**

Model answer

To make this investigation more reliable, you should repeat the investigation at least three times for each membrane thickness and then calculate a mean rate of diffusion of oxygen.

This is a model answer because it correctly applies a method of improving reliability to the specific example given in the question.

Command word: Write

Questions that ask you to 'Write' only require a short answer, not an explanation or a description. Usually, this command word will be used when the answer has to be written in a specific place, for example in a box or a table.

A Worked example

The following table gives the functions of two of the hormones found in plants. Write the names of the hormones in the appropriate boxes. **[2]**

Model answer

Function of hormone in plants	Hormone
Controls cell division and ripening of fruits	*Ethene*
Initiating seed germination	*Gibberellins*

This is a model answer because both hormones have been written correctly in the boxes.

Putting this into action!

Now that you know what all the main command words mean and how to answer them, the next and most important step is to put this learning into action. The next section provides some exam-style practice questions for you to apply your knowledge, and help you prepare for the exam. Do not forget that there are also past and sample assessment materials for your specific exam board online.

6 Exam-style questions

» Biology Paper 1

1 The graph below shows the changes in the number of chromosomes found in a single human cell during the process of meiosis.

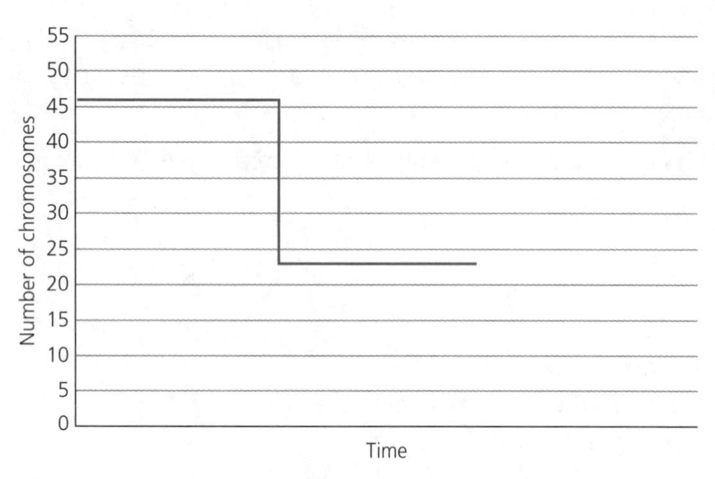

a State what type of cells are formed by meiosis. [1]

b Explain the shape of the graph. [2]

c Explain how the change in the number of chromosomes in the cell is important for the function of this cell. [2]

d Use the axes below to sketch a graph to show the chromosome number in a cell before and after mitosis has occurred. [2]

2 The table below shows the thicknesses of the walls of three different blood vessels.

Blood vessel	Thickness of wall (μm)
Artery	1500.00
Vein	700.00
Capillary	0.50

a Give the thickness of the artery in standard form, and to two significant figures. [2]

b Suggest how the thickness of each of the vessels is related to its function. [3]

c Identify which of the vessels in the table are affected by coronary heart disease, and explain the impact of this disease. [2]

3 The total volume of the alveoli in a human lung is $0.002\,m^3$, and the total surface area of the alveoli is $100\,m^2$.

a Give the surface area : volume ratio of the alveoli in the lung. [2]

b A human has an average surface area of $1.8\,m^2$ and a volume of $0.095\,m^3$. Calculate the surface area : volume ratio of the human. [2]

c Use these two values to explain why humans have an internal gas exchange surface. [3]

d The thickness of the alveoli and capillary wall is 2 μm. On a diagram of the alveoli, the thickness of the wall was 15 mm. Calculate the magnification of the diagram. [3]

e A student wanted to use Fick's law to calculate the rate of diffusion across the alveoli wall.

$$\text{Rate of diffusion} \propto \frac{\text{surface area} \times \text{concentration difference}}{\text{thickness of membrane}}$$

With the information provided in this question, would they be able to? Justify your answer. [3]

f In addition to the features listed above, give two other adaptations of the lungs for gas exchange. [2]

4 Interferon is a chemical that can be used to treat multiple sclerosis. Bacteria can be genetically engineered to produce interferon.

Explain the importance of the following enzymes in this process:

a DNA ligase [2]

b restriction enzymes. [2]

5 The speed of recovery for a heart rate to return back to its resting heart rate is a measure of fitness. Plan an investigation into the effect of the length of a period of exercise on the recovery of heart rate. [6]

6 An investigation was carried out into the effect of light intensity on the rate of photosynthesis. The results of the investigation are shown in the table below.

Distance from lamp (cm)	Rate of photosynthesis (bubbles / min)
20	20
40	15
60	8
80	3
100	2

a Plot the data on the axes below. This can be done on a separate piece of graph paper. [4]

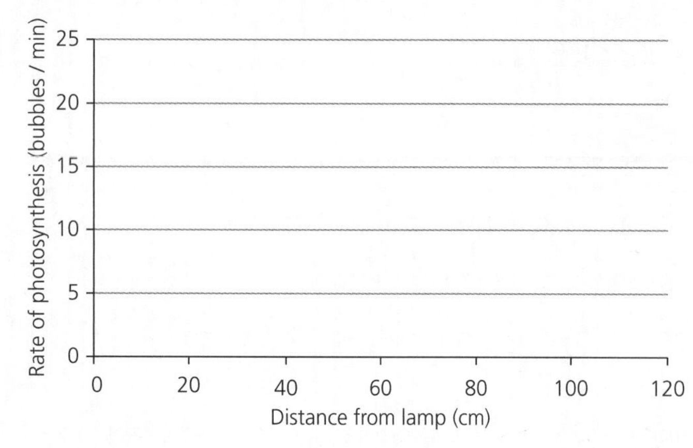

b The investigation was carried out at a constant temperature.

i Sketch a line on the graph to show the effect of increasing the temperature on the results. [1]

ii Explain the shape of the line you have drawn. [2]

7 The antibiotic linezolid blocks the transfer of amino acids to the ribosome by tRNA.

a Explain how linezolid prevents the growth of bacteria. [3]

b In an investigation into the effect of linezolid, the clear area below was produced on a bacterial lawn.

Use the formula below to calculate the clear area produced by linezolid. Give your answer to 1 d.p. [2]

$A = \pi r^2$

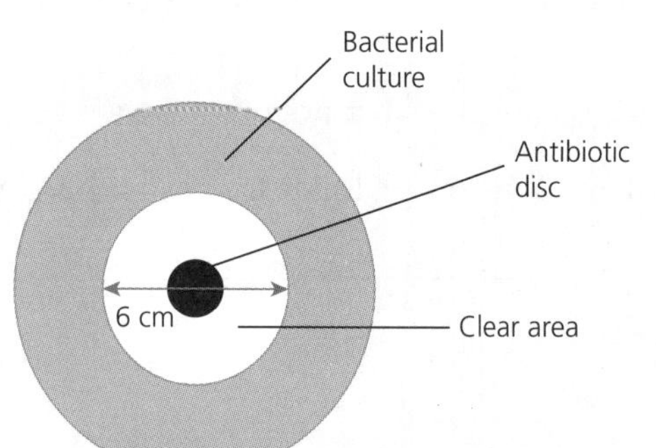

c Linezolid cannot be used to treat HIV. Explain why. [2]

d Linezolid was first synthesised by scientists in the lab. Compare this discovery to the discovery of penicillin. [2]

8 The blue mussel is a marine mollusc. It is a relative of other molluscs such as snails. It has the Latin name *Mytilus edulis*.

a Explain what information can be determined about the blue mussel from this name. [1]

b The table below shows the classification of the blue mussel. Complete the table. [3]

Kingdom	
	Mollusca
	Bivalvia
Order	Ostreoida

c Name the domain that the blue mussel is classified in, giving a reason for your answer. [2]

9 The diagram below shows the food web of an aquatic ecosystem.

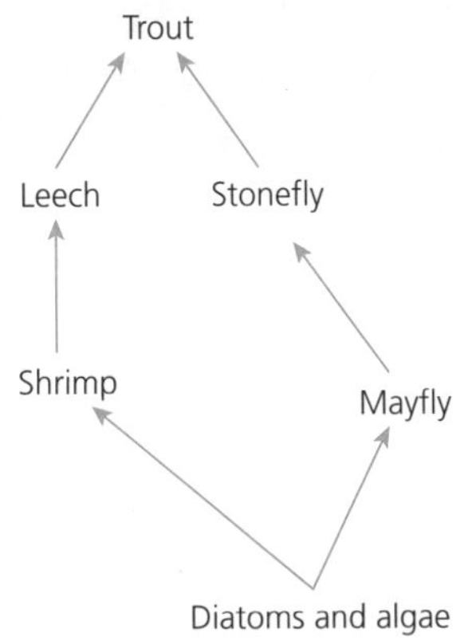

a Predict the effect of:

i a decline in the population of mayfly [2]

ii an increase in the population of trout. [2]

b Identify a:

i producer [1]

ii primary consumer [1]

iii secondary consumer. [1]

c A sampling activity found that there was an increase in sludge worms in the river. Suggest a possible reason for this increase. [2]

[Total = ... / 70 marks]

» Chemistry Paper 1

1 A student carried out an experiment to determine if the reaction between hydrochloric acid and sodium hydroxide was exothermic. The student followed the method below.

- Measure out 25.0 cm^3 of 0.10 mol/dm^3 hydrochloric acid and place in a polystyrene cup.
- Record the temperature of the hydrochloric acid.
- Gradually add 25.0 cm^3 of sodium hydroxide solution in 5.0 cm^3 portions to the hydrochloric acid, stirring after each addition.
- Record the temperature of the reaction mixture.

The table below shows the student's results.

Volume of sodium hydroxide added in cm^3	0.0	5.0	10.0	15.0	20.0	25.0
Temperature of reaction mixture in °C	20.5	21.5	22.5	23.5	25.2	28.0

a Plot a graph of the results. Use axes similar to those below. [3]

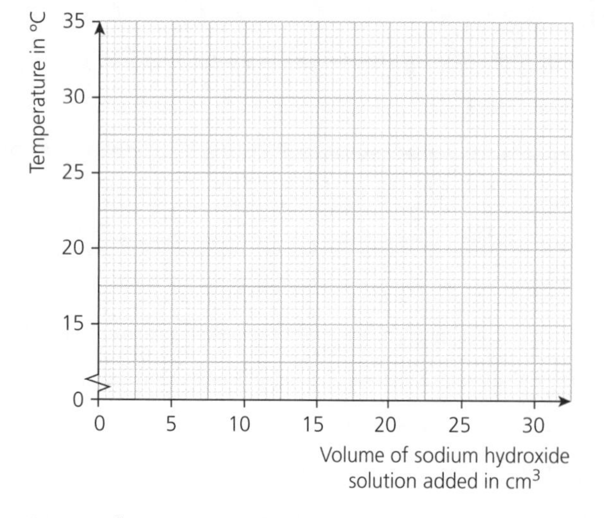

b State why your graph shows that this reaction was exothermic. [1]

c Name a piece of apparatus that could be used to add the sodium hydroxide solution to the acid. [1]

d Suggest one improvement that could be made to the apparatus used that would give more accurate results. Give a reason for your answer. [2]

e Write a balanced chemical equation for the reaction between sodium hydroxide and hydrochloric acid. [2]

f Calculate the number of moles of hydrochloric acid placed in the polystyrene cup. [1]

2 Bath crystals contain Epsom salts, which are hydrated magnesium sulfate crystals. Magnesium sulfate crystals can be prepared in the laboratory by reacting magnesium carbonate and sulfuric acid. The equation for the reaction is:

$MgCO_3 + H_2SO_4 \rightarrow MgSO_4 + H_2O + CO_2$

a State what could be is observed in this reaction. [1]

b Suggest one safety precaution that should be followed. [1]

c Calculate the maximum mass of magnesium sulfate that could be made when 2.1 g of magnesium carbonate is reacted with excess sulfuric acid. [3]

d The student obtained 1.8 g of magnesium sulfate. Calculate the percentage yield. [2]

e Suggest why the percentage yield is not 100% in this reaction. [1]

3 A student placed 25.0 cm^3 of white wine, containing tartaric acid, in a conical flask. The student carried out a titration to find the volume of 0.100 mol/dm^3 sodium hydroxide solution needed to neutralise the tartaric acid in the white wine.

a Name a suitable indicator for this titration and the colour change that would be seen. [2]

b Suggest why this titration is suitable for white wine, but it is not used to find the concentration of acid in red wine. [1]

c The student carried out four titrations. Her results are shown in the table below. Concordant results are within 0.10 cm^3 of each other.

Titration	Volume of 0.100 mol/dm^3 NaOH in cm^3
1	20.05
2	19.45
3	18.90
4	19.00

i Use the student's concordant results to work out the mean volume of 0.100 mol/dm^3 sodium hydroxide added. [2]

The equation for the reaction of tartaric acid in the white wine with the sodium hydroxide is:

$$C_4H_6O_6 + 2NaOH \rightarrow C_4H_4O_6Na_2 + 2H_2O$$

ii Calculate the concentration, in mol/dm^3, of the tartaric acid. Give your answer to two significant figures. [5]

iii Calculate the relative formula mass of tartaric acid ($C_4H_6O_6$). [1]

4 A student decides to investigate the reactivity of four metals, P, Q, R and S. Plan how the student could investigate the relative reactivity of the four metals, P, Q, R and S. The plan should use the fact that all four metals react exothermically with dilute sulfuric acid. You should name the apparatus used. [6]

5 A student investigated the reaction of 0.1 g of magnesium ribbon with 50 cm^3 of dilute hydrochloric acid of concentration 1 mol/dm^3 at 20 °C. The diagram below shows the apparatus used.

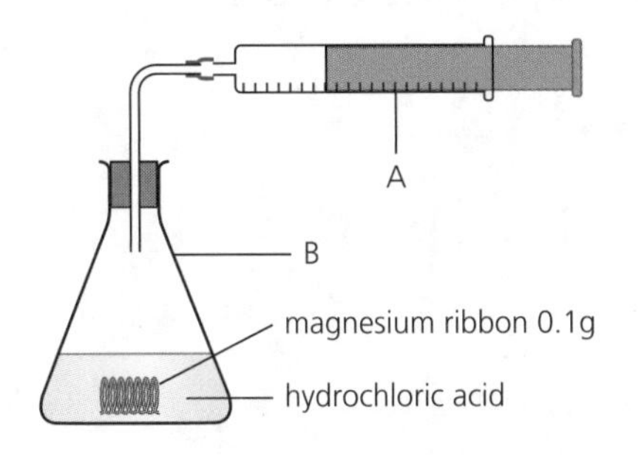

a Name the pieces of apparatus A and B. [2]

b Complete and balance the equation for the reaction between magnesium and hydrochloric acid:

.... + ... → ...H_2 +........................ [2]

c Give one advantage and one disadvantage of using a measuring cylinder to add the acid to the flask. [2]

d The table below shows the results of this experiment.

Time in s	0	30	60	90	120	150	180	210
Volume of gas in cm^3	0	13	22	30	36	43	49	49

Plot these results on the grid below and draw a line of best fit. [3]

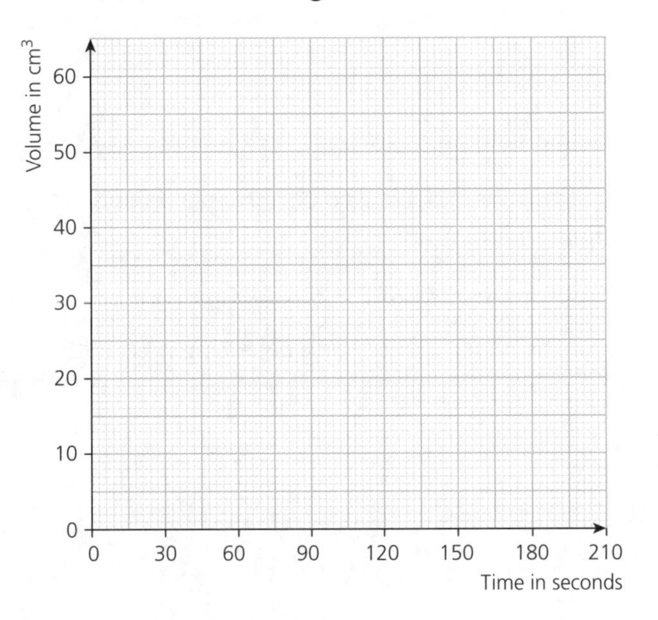

e Use your graph to find the time needed to collect 25 cm^3 of gas. [1]

f Using your graph, determine the rate of reaction at 60 seconds. Show your working on the graph. [4]

6 Solder is an alloy of tin and lead.

a A sample of a solder was made by mixing 22.5 g of lead with 15.0 g of tin. Calculate the percentage of tin by mass in this solder. [2]

b Why are alloys stronger than pure metals? Pick one option from below. [1]

A There are stronger bonds between the molecules they contain.

B They combine the properties of the metals from which they are made.

C They have atoms of different sizes in their structures.

D They are made using electrolysis.

7 The diagram shows the results of a chromatography experiment to analyse mixture of dyes.

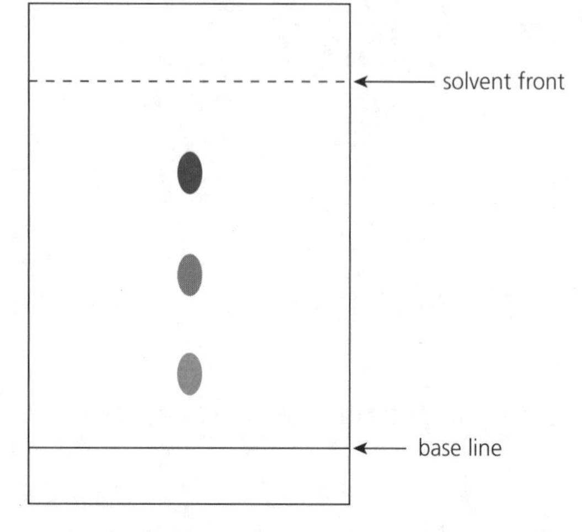

a Explain why the base line must be drawn in pencil instead of pen. [1]

b How many dyes were in the mixture? [1]

c Calculate the R_f value for the blue spot. Use the table below to identify it. [2]

Dye	R_f value
A	0.38
B	0.15
C	0.26
D	0.75
E	0.58

8 There is less carbon dioxide in the Earth's atmosphere now than there was in the Earth's early atmosphere.

The amount of carbon dioxide in the Earth's early atmosphere decreased because plants and algae used it for photosynthesis and it became locked up in sedimentary rocks.

a Photosynthesis can be represented by the equation shown below.

Complete the equation by writing the formula of the other product and balancing it correctly. [2]

.... CO_2 + ... H_2O → .... + ... O_2

b Explain what is meant by 'locked up carbon dioxide'. [2]

c The graph shows how the amount of carbon dioxide in the atmosphere has changed in recent years.

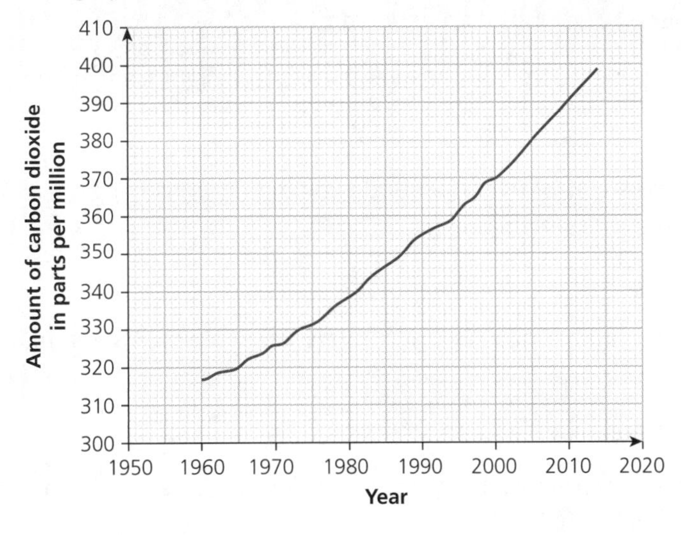

i Describe how the amount of carbon dioxide changed between 1960 and 2010. [1]

ii Calculate the percentage change in carbon dioxide levels between 2000 and 2010. [2]

iii Give two reasons why the amount of carbon dioxide has changed over time. [2]

9 Atmosphere is the term used to describe the collection of gases that surround a planet. The suggested composition of the atmosphere of Mars is shown in the table below.

Compare the composition of the Earth's atmosphere today with that of the planet Mars. [4]

Gas	Composition (%)
Carbon dioxide	95.0
Nitrogen	3.0
Noble gases	1.6
Oxygen	trace
Methane	trace

(Total marks .... / 70)

Physics Paper 1

1 The approximate diameter of an oil molecule can be measured by spreading a thin layer of oil over the surface of a large tray of water.

An oil-drop of volume 0.01 cm^3 forms a layer one molecule thick on the surface of a rectangular tray of water measuring 50 cm by 40 cm.

a Calculate the area of the water surface, giving your answer in m^2. [2]

b Calculate the volume of the oil drop in m^3. [1]

c Use your answers to parts **a** and **b** to calculate the diameter of the oil molecule. [2]

d Suggest whether your calculated diameter is likely to be an overestimate or an underestimate of the molecular diameter. [1]

2 a A student is asked to measure the thickness of a sheet of A4 paper. The student measures the thickness of a ream of the paper (500 sheets) and finds it to be 47 mm.

i Calculate the thickness of one sheet of paper. Give your answer in mm to 2 significant figures. [2]

ii The student measured the thickness of the ream to the nearest mm.

Calculate the minimum thickness of one sheet of the paper and explain your reasoning carefully. Give your answer in mm to 2 significant figures. [2]

b A metal wire is approximately 0.5 mm thick.

i Describe how the diameter of the wire could be measured accurately using a pencil and a ruler calibrated in mm. [3]

ii A student is given a long piece of the metal wire.

In addition to its diameter, suggest other measurements that the student needs to make in order to calculate the density of the metal. [2]

3 a Of the chemical energy used in a car's engine, $\frac{7}{10}$ is converted into heat.

$\frac{9}{10}$ of the remainder is converted into useful energy.

Petrol contains 32 MJ per litre.

A motorist fills her car's petrol tank with 40 litres of fuel.

i Calculate the total energy content of the 40 litres of petrol. [2]

ii Calculate how much of this energy will eventually be converted into heat. [1]

iii Calculate how much energy will be converted into useful energy. [1]

iv Use your answers to parts **i** and **iii** to calculate the efficiency of the car's engine. [2]

b The car in part **a** travels 350 km on a tank containing 20 litres of petrol. A car manufacturer claims that its electric car can go exactly the same distance on a single battery charge of 150 MJ.

i Calculate how much more energy the petrol car uses than the electric car to travel this distance. [2]

ii Explain why you cannot tell from this data alone that the electric car is more efficient. [1]

Governments are encouraging more people to use electric cars, partly because they claim they are better for the environment.

iii Evaluate the advantages and disadvantages of using electric cars rather than petrol cars. [6]

4 In a computer simulation, the predicted activity of a particular short-lived radioisotope is tabulated every 30 minutes. The results are shown in this table.

Activity (Bq)	600	476	378	300	238	189	150	119	94	75
Time (hours)	0	0.5	1.0	1.5	2.0	2.5	3.0	3.5	4.0	4.5

a Calculate the probability that a particular nucleus will decay in a given period of 1 hour. [2]

b Plot the graph of *activity* (*y*-axis) against *time* (*x*-axis) and draw a line of best fit. [6]

c From your graph, show that the half-life of this radioisotope is approximately 1.5 hours. [1]

d Estimate the activity of this radioisotope after 1.8 hours. [2]

5 A current of 480 mA enters a network of resistors as shown in the diagram.

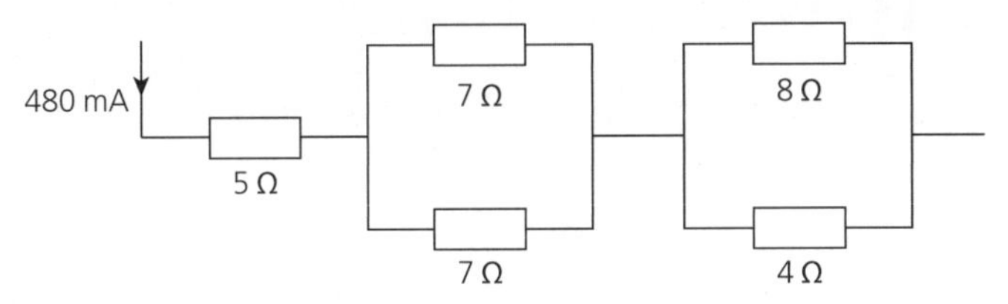

a Complete the table to show the current in each resistor. [4]

Resistance (Ω)	5	7	8	4
Current (mA)				

b Across which resistor is the voltage the greatest? Justify your answer. [3]

c Predict what would happen to the current in the 8 Ω resistor if the current in each of the 7 Ω resistors was doubled. Justify your answer. [2]

6 **a** The diagram shows a transformer, designed to decrease a voltage from 12 kV to 24 V. Coil B has 25 000 turns.

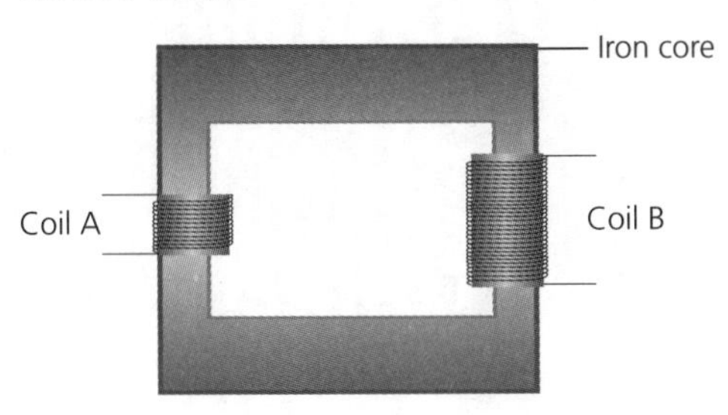

i Identify which coil, A or B, is the primary coil. Justify your answer. [3]

ii Calculate the turns ratio of this transformer and use it to find the number of turns in coil A. [3]

b This diagram represents an electricity transmission system, such as the National Grid.

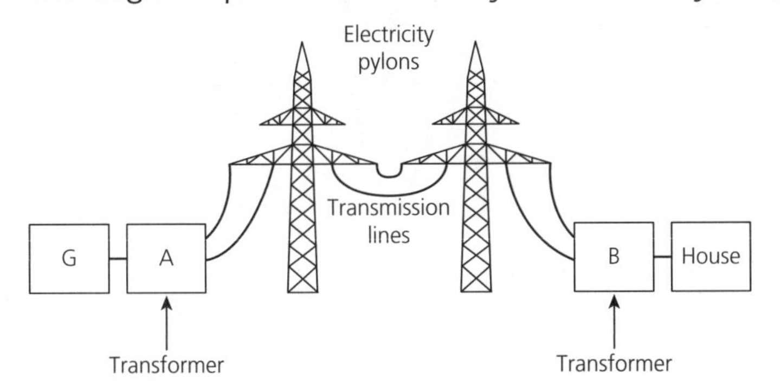

i State what is represented by box G. [1]

ii Identify which of the transformers, A or B, is the step-up transformer. [1]

In the Snowdonia National Park in Wales there are no electricity pylons to be seen, but there is an electrical distribution system.

iii Suggest how the electricity is distributed in Snowdonia, and suggest why this method is not used throughout the country. [3]

7 Nuclear attack submarines are capable of diving to a depth where the water pressure on the hull is 7.35×10^6 N/m^2.

a **i** If the average density of seawater is 1050 kg/m^3 (1.05 g/cm^3), calculate the depth to which such a submarine can submerge. [3]

ii Suggest a reason why the total pressure on the hull at this depth is greater than 7.35×10^6 N/m^2. [1]

b A bubble of volume 0.1 cm^3 is released in error by the submarine and the bubble rises to the surface. The pressure on the bubble at the moment of release is 6×10^6 N/m^2.

i Explain why the volume of the bubble increases as it rises to the surface. [1]

ii Estimate the volume of the bubble just as it reaches the surface if the pressure on it there is 1×10^5 N/m^2. [3]

The temperature of the surrounding water at the moment the bubble is released is generally lower than the temperature at the surface.

iii If the change in temperature was considered, identify whether it would lead to an increase or decrease in the volume estimated in your answer to part **ii**. [1]

[Total = ... / 70 marks]

Answers

» Maths

Units

Converting between units (pages 2–4)

Guided questions

1 $1.2 \times 1000 = 1200\,cm^3$

2 **Step 1** $8.2 \times 1000 = 8200\,kg$

Step 2 $8200 \times 1000 = 8\,200\,000\,g$

Practice question

3 a To convert from dm^3 to cm^3 multiply by 1000

$1.2 \times 1000 = 1200\,cm^3$

b To convert from cm^3 to dm^3 divide by 1000

$\frac{420}{1000} = 0.42\,dm^3$

c To convert from cm^3 to dm^3 divide by 1000

$\frac{3452}{1000} = 3.452\,dm^3$

d To convert from tonnes to grams, first convert to kg by multiplying by 1000 and then convert to grams by multiplying by 1000 (or multiply by 10^6 in one step)

$4.4 \times 1000 \times 1000 = 4\,400\,000\,g$

e To convert from kg to g multiply by 1000

$4 \times 1000 = 4000\,g$

f To convert from g to kg divide by 1000

$\frac{3512}{1000} = 3.512\,kg$

Calculations that often involve conversion of units (pages 4–6)

Guided questions

1 **Step 1** $9.8 \times 1000 = 9800\,g$

Step 2

$$\text{amount (in moles)} = \frac{\text{mass (g)}}{M_r} = \frac{9800}{98} = 100\,\text{mol}$$

2 **Step 1** $\frac{48\,000}{1000} = 48\,dm^3$

Step 2 Substitute the volume into the equation and calculate your final answer.

$$\text{amount (in moles)} = \frac{\text{volume}(dm^3)}{24} = \frac{48}{24} = 2\,\text{mol}$$

Practice questions

3 The mass of calcium must be converted from kg to g before calculating moles.

$6 \times 1000 = 6000\,g$

$$\text{amount (in moles)} = \frac{\text{mass (g)}}{A_r} = \frac{6000}{40} = 150\,\text{mol}$$

4 The mass of calcium carbonate must be converted from tonnes to grams before calculating moles.

$3.2 \times 1000 \times 1000 = 3\,200\,000\,g$

$$\text{amount (in moles)} = \frac{\text{mass(g)}}{M_r} = \frac{3\,200\,000}{100} = 32\,000\,\text{mol}$$

5 a The mass of ammonia must be converted from kg to g by multiplying by 1000.

mass of ammonia in grams $= 17 \times 1000 = 17\,000\,g$

b Calculate the relative formula mass of $NH_3 = 14 + (3 \times 1) = 17$

$$\text{amount (in moles)} = \frac{\text{mass (g)}}{M_r} = \frac{17\,000}{17} = 1000\,\text{mol}$$

6 Mass of iron(III) oxide in grams $= 2.1\text{ tonnes} \times 1000 \times 1000 = 2\,100\,000\,g$

Relative formula mass $= (2 \times 56) + (3 \times 16) = 160$

$$\text{amount (in moles)} = \frac{\text{mass (g)}}{M_r} = \frac{2\,100\,000}{160} = 13\,125\,\text{mol}$$

7 Mass of magnesium nitrate in kilograms $= 0.592 \times 1000 = 592\,g$

Relative formula mass $- 24 + (2 \times 14) + (6 \times 16) = 148$

$$\text{amount (in moles)} = \frac{\text{mass (g)}}{M_r} = \frac{592}{148} = 4\,\text{mol}$$

8 Volume of sulfur trioxide in $dm^3 = \frac{7200}{1000} = 7.2$

$$\text{amount (in moles)} = \frac{\text{volume}\,(dm^3)}{24} = \frac{7.2}{24} = 0.3\,\text{mol}$$

Arithmetical and numerical computation

Expressions in decimal form (pages 6–8)

Guided questions

1 Root diameter to one decimal place = 0.3 cm

2 $\frac{20\,cm}{6.4\,s} = 3.125\,cm/s$

speed of car = 3.13 cm/s

Practice questions

3 The cathode increases in mass and so copper is deposited here.

Mass $= 1.87 - 1.58 = 0.29\,g = 0.3$ (to 1 d.p.)

4 Formula for pressure:

$$P = \frac{F}{A}$$

$$= \frac{630}{205}$$

$$= 3.07\,N/cm^2$$

$$= 3.1\,N/cm^2 (1\,d.p.)$$

Standard form

Powers of 10 (pages 8–10)

Guided questions

1 **Step 1** Amount (in moles) $= \frac{\text{mass (g)}}{M_r} = \frac{2.3}{23} = 0.1$

Step 2 $0.1 \times 6.02 \times 10^{23} = 6.02 \times 10^{22}$

2 **Step 1** Bacterial population $= 10 \times 2^{12} = 10 \times 4096$

Step 2 Bacterial population $= 40\,960 = 4.096 \times 10^4$

Practice questions

3 distance = speed × time

$= 25$ mm per year × 500 000 years

$= 1.25 \times 10^7$ mm

$= 1.25 \times 10^4$ m

4 Bacterial population =

initial bacterial population $\times 2^{\text{number of divisions}}$

Number of divisions $= 30 \div 5 = 6$

$200 \times 2^6 = 12\,800$

$12\,800 = 1.28 \times 10^4$

5 a atom $= 0.256 \times 10^{-9}$ m $= 2.56 \times 10^{-10}$ m,

wire $= 0.044$ cm $= 0.044 \times 10^{-2}$ m $= 4.4 \times 10^{-4}$ m

b $\frac{4.4 \times 10^{-4}}{2.56 \times 10^{-10}} = 1\,718\,750 = 1.72 \times 10^6$ (to 2 d.p.)

6 1.67×10^{-24}

Ratios, fractions and percentages

Fractions (pages 11–14)

Practice questions

1 a $1\frac{1}{4} + 3\frac{5}{8} = 4\frac{7}{8}$

b $2\frac{2}{3} + 4\frac{5}{6} = 7\frac{1}{2}$

c $7\frac{5}{12} - 6\frac{1}{4} = 1\frac{1}{6}$

d $3\frac{2}{5} - 4\frac{7}{10} = -1\frac{3}{10}$

2 $\frac{5}{8}$ of the gold is copper

$\frac{1}{8}$ of the gold $= \frac{1}{5}$ of 95 = 19 grams

All the gold $= \frac{8}{8} = 8 \times 19 = 152$ grams

Percentages (pages 14–16)

Guided questions

1 **Step 1** $M_r = 40 + (14 \times 2) + (16 \times 6) = 164$

Step 2 $14 \times 2 = 28$

Step 3 $\frac{\text{mass of nitrogen}}{M_r} = \frac{24}{164}$

Step 4 $\frac{248}{164} \times 100 = 17\%$

2 **Step 1** Efficiency of energy transfer $= 52 \div 4000 \times 100$

$= 0.013 \times 100$

Step 2 Efficiency of energy transfer = 1.3%

Practice questions

3 a percentage wasted $= \frac{\text{wasted energy}}{\text{total input energy}} \times 100\%$

$= \frac{21 \text{ MJ}}{30 \text{ MJ}} \times 100\%$

$= 70\%$

b percentage usefully converted =
100% – percentage wasted = 100% – 70% = 30%

4 a Energy in heather = 300 000 kJ

Energy in grouse = 19 000 kJ

As a fraction $= \frac{19\,000}{300\,000}$

$= \frac{19000 \div 1000}{300000 \div 1000}$ as 1000 is the largest common factor $= \frac{19}{300}$

As a percentage $= \frac{19\,000}{300\,000} \times 100 = 6.3\%$

b Energy in grouse = 19 000 kJ

Energy in fox = 2100 kJ

As a fraction $= \frac{2100}{19\,000}$

$= \frac{2100 \div 100}{19\,000 \div 100}$ as 100 is the largest common factor

$= \frac{21}{190}$

As a percentage $= \frac{2100}{19\,000} \times 100 = 11\%$

5 a $\frac{2 \times 1}{74} \times 100 = 2.7\%$

b $\frac{2 \times 39}{294} \times 100 = 26.6\%$

c $\frac{2 \times 14}{132} \times 100 = 21.2\%$

Ratios (pages 16–18)

Guided questions

1 **Step 1**

P : O

0.050 : 0.125

Step 2

$\frac{0.050}{0.050} : \frac{0.125}{0.050}$

1 : 2.5

$1 \times 2 : 2.5 \times 2$

2 : 5

P_2O_5

2

	r	r
R	Rr	Rr
r	rr	rr

- Expected ratio = 2 Rr : 2 rr = 1 Rr : 1 rr
- Therefore expected phenotype ratio = 1 red stripe : 1 orange stripe

Practice questions

3

Voltage, *V* (V)	3.2	4.0	4.8	5.6	6.4	7.2
Current, *I* (A)	0.20	0.25	0.30	0.35	0.40	0.45
Ratio	16:1	16:1	16:1	16:1	16:1	16:1

- Voltage is directly proportional to the current since ratio $V:I$ is constant

4 a $C_4H_5N_2O$

b $Na_2S_2O_3$

c CH_2O

d P_2O_5

Balancing equations (pages 18–20)

Guided question

1 **Step 1** $N_2 : H_2$

1 : 3

Step 2 There is three times as much H_2 as N_2, so divide H_2 moles by 3

$$\frac{0.4}{3} = 0.13$$

Practice question

2 a $Cu(NO_3)_2 : O_2$

2 : 1

4 : 2 mol

b $Cu(NO_3)_2 : 4NO_2$

2 : 4

1 : 2

0.6 : 0.6×2=0.12

Estimating results (pages 20–21)

Guided question

1 **Step 2** Compound distance=8 solvent distance=20

Step 3 Estimate the R_f value

$$\frac{8}{20} = \frac{4}{10} = 0.4$$

The R_f is approximately 0.4 so the compound is P

Practice questions

2 estimated distance there and back = 400 000 + 400 000 = 800 000 km = 800 000 000 m

$$\text{estimated time} = \frac{\text{distance}}{\text{speed}} = \frac{800\,000\,000 \text{ m}}{3 \times 10^8 \text{ m/s}}$$

= 3 s (to nearest second)

3 No, it is not the best estimate. A better estimate would be to round 3.9 to 4 rather than 3. This would instead give an estimate of:

$$\frac{4}{6} \times 100\% = 67\% \text{ (to 2 d.p.)}$$

Using sin and sin^{-1} keys (pages 21–23)

Guided question

1 **Step 1** $n = \frac{1}{\sin c} \Rightarrow c = \sin^{-1}\left(\frac{1}{1.52}\right)$

Step 2 $c = 41.1395° = 41.1°$ (1 d.p.)

Practice questions

2 refractive index $= \frac{\sin 90}{\sin 40} = \frac{1}{0.6428} = 1.56$ (2 d.p.)

3 refractive index $= \frac{\sin 60}{\sin(90-60)} = \frac{0.866}{0.500} = 1.73$ (2 d.p.)

Handling data

Significant figures (pages 23–25)

Guided question

1 **Step 1** $V = I \times R$

Step 2 $V = 1.4 \times 6.8$

Step 3 $V = 9.52$ volts

Step 4 Number of s.f. in data in question is 2.

Step 5 $V = 9.5$ volts (2 s.f.)

Practice questions

2 As the measurements are both made to 3 s.f., this answer should also be given to 3 s.f.

Therefore: 6.819 g/hour = 6.82 g/hour (to 3 s.f.)

3 $E = mc\Delta\theta$

= 2.55 kg × 4200 J/kg °C × 12.2 °C

= 130 662 J

But the number of significant figures in the question is 3, so:

E = 131 000 J (3 s.f.)

4 $\% = \frac{2.53}{2.85} \times 100 = 88.8819 = 89\%$ (to 2 s.f.)

Finding arithmetic means (pages 26–27)

Guided question

1 a **Step 1** The outlier is 0.5 Ω.

b **Step 1** The sum of the other results is: 2.1 + 2.2 + 1.9 + 1.8 = 8

Step 2 The mean resistance is 8 ÷ 4 = 2 Ω

Practice questions

2 Mean of B = $(1500 + 1600 + 1700) \div 3$

Mean of B = $4800 \div 3 = 1600$

3 River A: $\frac{14 + 13 + 11 + 9 + 8}{5} = 11$

River B: $\frac{8 + 9 + 10 + 11 + 9}{5} = 9.4$

River B is safest

4 Mean of 10 values = 4.2

So, sum of these 10 values is $4.2 \times 10 = 42$

Sum of the given 9 values = $(4.1 + 4.2 + 4.2 + 4.3 + 4.3 + 4.1 + 4.2 + 4.0 + 4.1) = 37.5$

So, missing value is $42 - 37.5 = 4.5$ J/g°C

Calculating weighted means (pages 27–28)

Guided question

1 **Step 1** $79 + 10 + 11 = 100$

Step 2

$$\text{relative atomic mass} = \frac{(79 \times 24) + (10 \times 25) + (11 \times 26)}{100} = 24.32 = 24.3$$

Practice questions

2 $\frac{(69 \times 63) + (31 \times 65)}{100} = 63.62 = 63.6$

3 $\frac{(95.02 \times 32) + (0.76 \times 33) + (4.22 \times 34)}{100} = 32.09$

Constructing frequency tables, bar charts and histograms

Frequency tables (pages 29–30)

Practice question

1 a

Number on die	Tally	Frequency
1	𝍸 ||||	9
2	𝍸 ||||	9
3	𝍸 𝍸 ||	12
4	𝍸 𝍸 ||	12
5	𝍸 |||	8
6	𝍸 𝍸	10
	Total	60

b We would expect an unbiased die to show about 10 of each of the numbers 1–6.

No number comes up more than 12 times or fewer than 8 times. Based on this distribution, there is no strong evidence of bias.

Bar charts (pages 32–34)

Guided question

1 a **Step 1** Female students have a red-coloured bar.

Step 2 This bar is biggest in 2016.

b **Step 1** Year in which difference in males and females is greatest is the year in which there is greatest difference in the height of the bars.

Step 2 This year is 2016.

c **Step 1** Number of male students in 2015 = 80

Step 2 Number of female students in 2015 = 66

Step 3 Total number of students in 2015 = 80 + 66 = 146

Practice questions

2

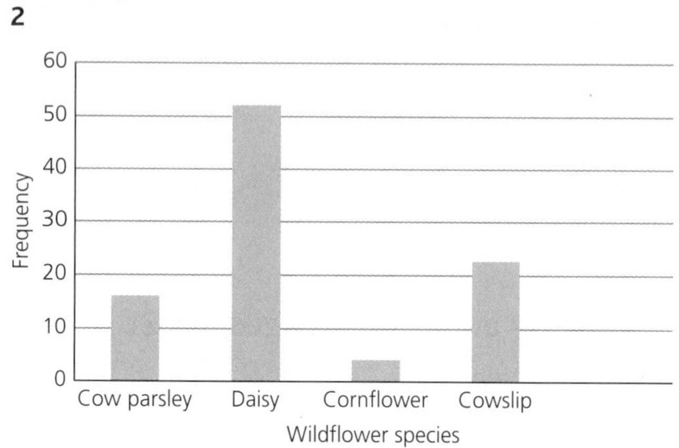

Bar chart showing the frequency of wildflower species

3 a

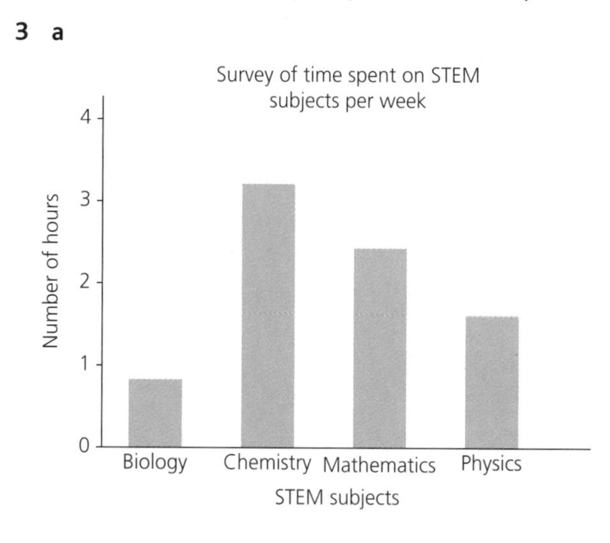

b Physics – Biology = $1.6 - 0.8 = 0.8$ hours

c Total time = $(0.8 + 3.3 + 2.4 + 1.6) = 8.1$ hours

Histograms (pages 35–36)

Guided question

1 **Step 1** The number of days when the snowfall was between 30 and 40 mm was 6.

Step 2 The numbers missing from the middle column of the table are 25, 35, 45 and 55.

Step 3 The vertical axis is labelled *Number of days*. The horizontal axis is labelled *Snowfall* (mm) and will range from 0 to 60.

Step 4 The first bar is centred on 15 mm, three days in height and 10 mm wide.

Step 5 Draw the remaining bars. The final bar is two days in height, centred on 55 mm and 10 mm wide.

Step 6 Add the title to the histogram.

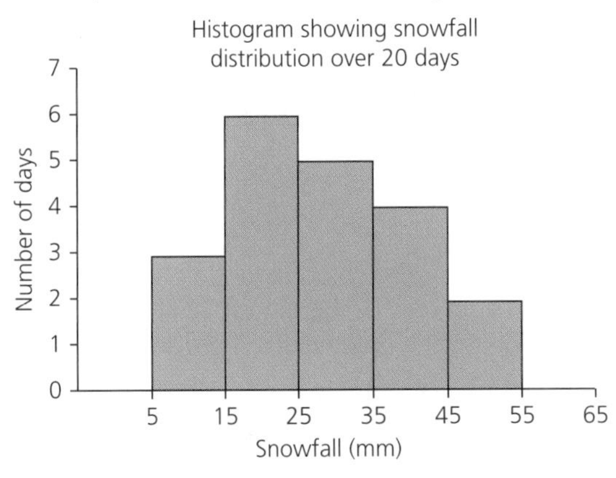

Practice question

2

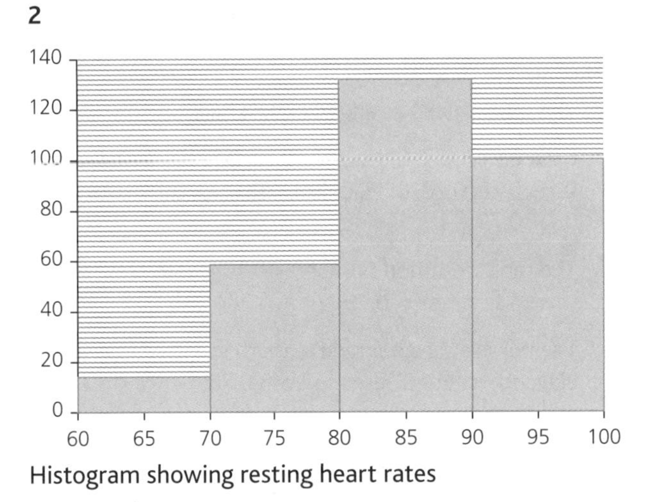

Histogram showing resting heart rates

Pie charts (pages 36–38)

Guided question

1 **Step 1** The number of students surveyed altogether was: 90.

Step 2 Each student in the pie chart is represented by an angle of 4 degrees.

Step 3 So, the angles for each method of transport are:

Walking = 60°; Cycling = 20°; Car = 140°; Bus = 80°; Train = 60°

Step 4 With a compass, draw a large circle to represent the pie.

Step 5 With a ruler, draw a line from the centre of the circle to its circumference.

Step 6 Draw the sectors in the pie using the angles found in Step 3, then label the sectors.

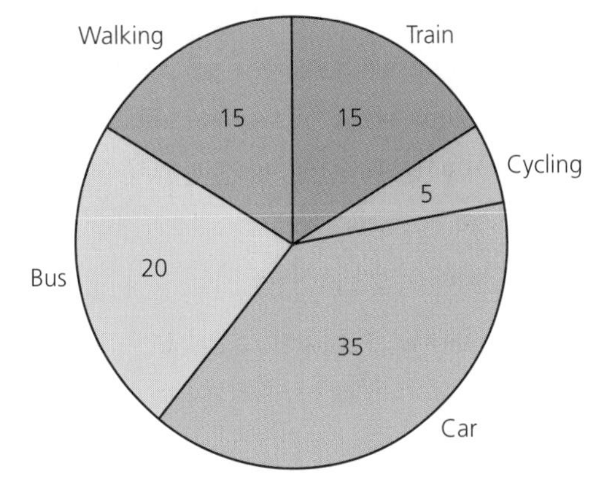

Practice question

2 a Number of people using wood =
$180 - (90 + 45 + 25 + 10 + 2) = 180 - 172 = 8$

b There are 180 people, so each person represents 2°

Energy resource	Gas	Oil	Coal	Electricity	Wood	Other
Number of people	90	45	25	10	8	2
Sector angle (degrees)	180	90	50	20	16	4

c

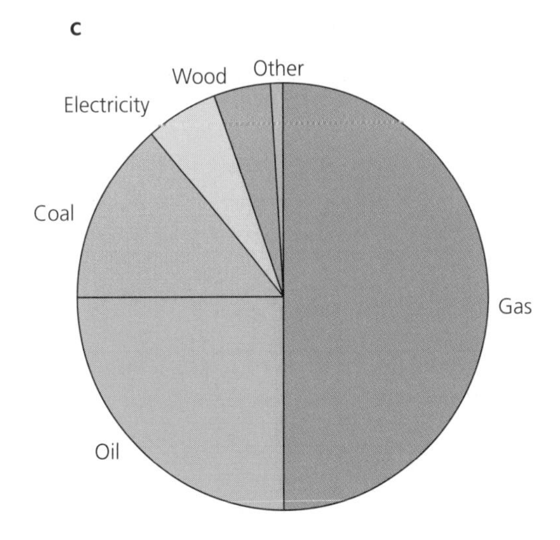

Understanding the principles of sampling (pages 38–39)

Guided questions

1 Population size = (total number in first sample × total number in second sample) ÷ number marked in second sample

Step 2 Population size $= (92 \times 78) \div 15$
$= 478$ (rounded to the nearest whole number)

2 **Step 1** 7 quadrants contain *Digitalis* out of a total of 10 quadrants.

Step 2 Therefore, species frequency $= (7 \div 10) \times 100$
$= 70\%$

Practice questions

3 Number of snails = (total number in first sample × total number in second sample) ÷ number marked in second sample

Number of snails = (105 × 120) ÷ 45

Number of snails = 12 600 ÷ 45 = 280

4 % cover of grass = number of squares containing grass ÷ total number of squares

% cover of grass = (15 ÷ 25) × 100 = 60%

Simple probability (pages 39–42)

Guided question

1 a **Step 1** In one half-life the number of undecayed nuclei falls by 50%.

Step 2 So, in one half-life the 80 million undecayed nuclei will fall to 40 million.

Step 3 From the graph this takes 2 minutes.

b **Step 1** In 6 minutes the number of undecayed nuclei has fallen to 10 million.

Step 2 So, the fraction that have decayed is $\frac{7}{8}$.

Step 3 So, the probability of decay within 6 minutes is 0.875.

Practice questions

2 a

Time elapsed (mins)	0	1	2	3	4	5	6	7
Expected number of undecayed nuclei	3000	2100	1470	1029	720	504	353	247

b and c

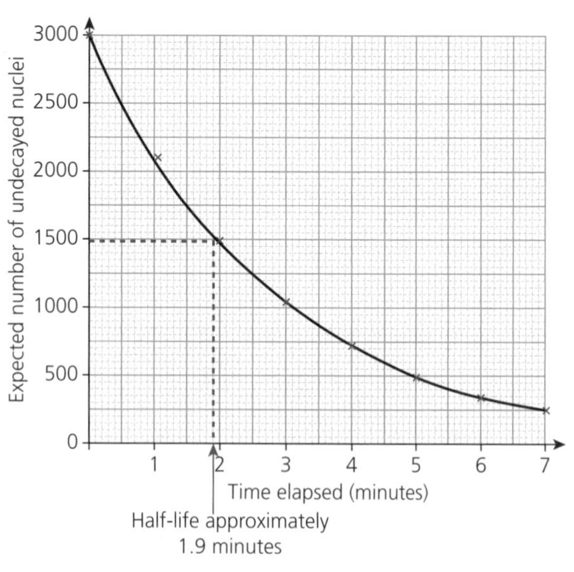

3 Parents: Gg Gg

Gametes: G g G g

	G	g
G	GG	Gg
g	Gg	gg

Offspring 25% GG, 50% Gg, 25% gg

As only offspring with genotype gg will have yellow fruit, the probability of an offspring having yellow fruit is 25%.

Understanding mean, mode and median (pages 42–43)

Guided question

1 **Step 1** 1, 2, 3, 4, 6, 7, 9

Step 2 Since the mode is 9, there must be at least two 9s. So, add 9 to the ordered list. There are now 8 numbers in the ordered list.

Step 3 The median is the fifth number in the ordered list, so the missing number must be 6, 7, 8 or 9. Since the mode is 9, the missing number cannot be 6 or 7. Therefore, the missing number must be 8 or 9.

Practice question

2 a The 20 numbers arranged in order are:

0.77, 0.77, 0.77, 0.77, 0.77, 0.78, 0.78, 0.78, 0.78, 0.78, 0.78, 0.78, 0.79, 0.79, 0.79, 0.79, 0.79, 0.79, 0.79, 0.79

The most common number is 0.79 (there are eight of them), so mode = 0.79

b The numbers in the middle are the tenth and eleventh values. These are both 0.78.

So, median = 0.78

Using a scatter diagram to identify a correlation (pages 44–46)

Guided question

1 **Step 1** As the distance from the tree increases, the percentage cover of grass increases.

Step 2 This shows a positive correlation.

Practice question

2

Distance *D* of molecule from a fixed point (mm)	0	15	21	26	30	33	38	40
Collision number *N*	0	1	2	3	4	5	6	7
$\sqrt{N}$	0	**1.0**	1.4	**1.7**	2.0	**2.2**	**2.4**	**2.6**

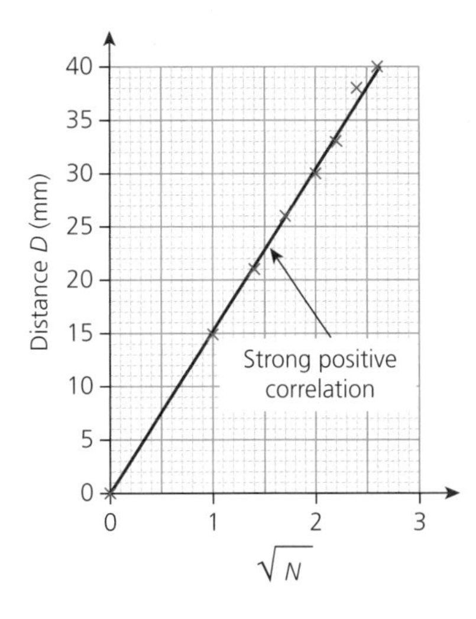

Orders of magnitude (pages 46–47)

Guided question

1 **Step 1** electrical force ÷ gravitational force
$= 9.22 \times 10^{-8}$ N ÷ 4.06×10^{-47} N $= 2.27 \times 10^{39}$

Step 2 The electrical force is 10^{39} times greater than the gravitational force.

Practice questions

2 $\frac{1 \times 10^{-4}}{1 \times 10^{-9}} = 1 \times 10^{5} = 100\,000$ times bigger

3 $\frac{1.0 \times 10^{-6}}{1.6 \times 10^{-9}} = 0.625 \times 10^{3} = 625$ times bigger

Using the magnification equation (pages 47–48)

Guided question

1 **Step 1** Image size = 150; object size = 2

Step 2 Magnification = 150 ÷ 2 = 75

Practice question

2 Magnification = image size ÷ object size

Object size = image size ÷ magnification

Object size = 30 ÷ 340 = 0.088235 = 0.088 mm (2 s.f.)

Algebra

Understanding and using algebraic symbols (pages 50–52)

Guided question

1 Rate of decomposition ∝ soil temperature

Practice questions

2 Blood pressure in arteries > blood pressure in veins

3 Enzyme concentration ∝ rate of reaction

Rearranging the subject of an equation (pages 52–54)

Guided questions

1 **Step 1** $2E = kx^2$

Step 2 $\frac{2E}{k} = \frac{kx^2}{k}$

Step 3 $\frac{2E}{k} = x^2$

Step 4 $\sqrt{\frac{2E}{k}} = x$

Step 5 $x = \sqrt{\frac{2E}{k}}$

2 **Step 2**

$$\frac{(\text{volume} \times \text{conc.} \times \cancel{1000})}{\cancel{1000}} = \text{moles} \times 1000$$

$$\frac{\text{volume} \times \cancel{\text{conc.}}}{\cancel{\text{conc.}}} = \frac{\text{moles} \times 1000}{\text{conc.}}$$

$$\text{volume} = \frac{\text{moles} \times 1000}{\text{conc.}}$$

Practice questions

3 a $x = \frac{y-1}{2}$

b $x = \frac{4+y}{3}$

c $x = \frac{y-c}{m}$

d $x = 1 - 2y$

4 a Theoretical yield $= \frac{\text{actual yield} \times 100}{\text{percentage yield}}$

b Conc. $= \frac{\text{volume} \times 1000}{\text{moles}}$

c Time taken $= \frac{\text{quantity of reactant used}}{\text{mean rate of equation}}$

Substituting values into an equation (pages 54–56)

Guided questions

1 **Step 3** volume × conc. = moles × 1000

Step 4 conc. $= \frac{\text{moles} \times 1000}{\text{volume}}$

Step 5 conc. $= \frac{0.0034 \times 1000}{15.0} = 0.23\,\text{mol/dm}^3$

2 **Step 1** $a = \frac{v-u}{t}$

Step 2 $a = \frac{30-0}{12}$

Step 3 $a = 2.5\,\text{m/s}^2$

Practice question

3 Area of clear zone $= \pi r^2 = \pi \times 72^2$

Area of clear zone $= \pi \times 518$

Area of clear zone $= 16\,286\,\text{mm}^2$

Solving simple equations (page 56)

Practice questions

1 $R_f = \frac{\text{distance moved by substance}}{\text{distance moved by solvent}}$

Distance moved by solvent × R_f = distance moved by substance

$10.2 \times 0.80 = 8.2\,\text{cm}$

2

$$\text{atom economy} = \frac{\text{sum of relative formula mass of desired product from equation}}{\text{sum of relative masses of all reactants from equation}} \times 100$$

$$= \frac{160}{124 + 98} \times 100 = 72\%$$

3 $v = \lambda$

$3 \times 10^8 = 5 \times 10^{14} \times \lambda$

$$\lambda = \frac{3 \times 10^8}{5 \times 10^{14}}$$

$= 6 \times 10^{-7}\,\text{m}$

4 Energy available to primary consumers = energy in primary producers − energy lost in respiration − energy lost by waste and death

Energy in primary producers = energy available in primary producers + energy lost in respiration + energy lost by waste and death

Energy in primary producers = 20 000 + 30 000 + 150 000

Energy in primary producers = 200 000 kJ

Inverse proportion (pages 56–58)

Guided question

1 **Step 1** PR = a constant

Step 2 $PR = 1200 \times 48 = 57\,600$

Step 3 $57\,600 = P \times 60$

Step 4 $P = \frac{57\,600}{60} = 960\,\text{W}$

Practice question

2 Since I and d^2 are inversely proportional, the product $I \times d^2$ is a constant.

In this case $I \times d^2 = 1440$

For the ship: $0.001 \times d^2 = 1440$

a $d^2 = \frac{1440}{0.001} = 1\,440\,000$

b $d = \sqrt{1\,440\,000} = 1200\,\text{m}$

Graphs

Translating between graphical and numerical form (pages 60–62)

Guided question

1 a **Step 1** The speed on the vertical axis is half way between 6 m/s and 8 m/s.

Step 2 At this speed, draw a horizontal line to the graph.

Step 3 From the point where this line meets the graph, draw a vertical line to the time axis.

Step 4 The line meets the time axis at 2.5 seconds. This is the answer.

b **Step 1** The time on the horizontal axis is half way between 1 s and 2 s.

Step 2 At this speed, draw a vertical line to the graph.

Step 3 From the point where this line meets the graph, draw a horizontal line to the vertical axis.

Step 4 The line meets the speed axis at 9 m/s. This is the answer.

Practice question

2 a Result at 4.5 minutes

b Mass of flask and contents /g; time /s

c A graph of mass of flask and contents against time for the reaction of calcium carbonate and acid

d Yes, as it fits most of the graph paper

Understanding that $y = mx + c$ represents a linear relationship (pages 62–64)

Guided question

1 **Step 1** $m = -0.5$; $c = 9$

Step 2 At $x = 0$, $y = 9$

Step 3 At $x = 10$, $y = 4$

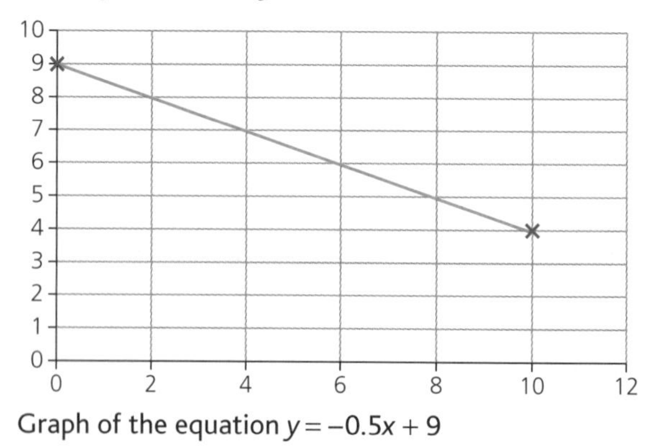

Graph of the equation $y = -0.5x + 9$

Practice question

2

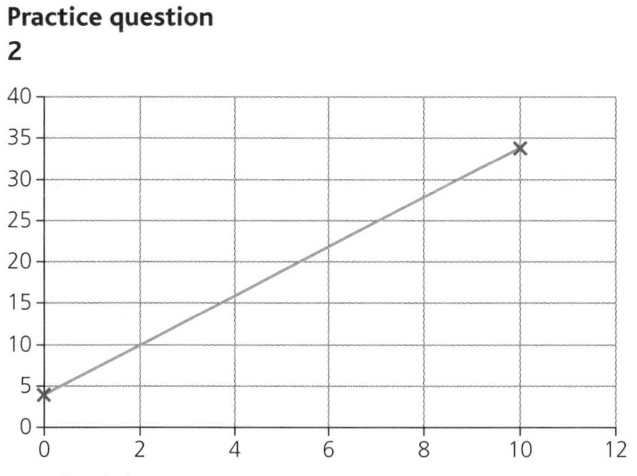

Graph of the equation $y = 3x + 4$

4

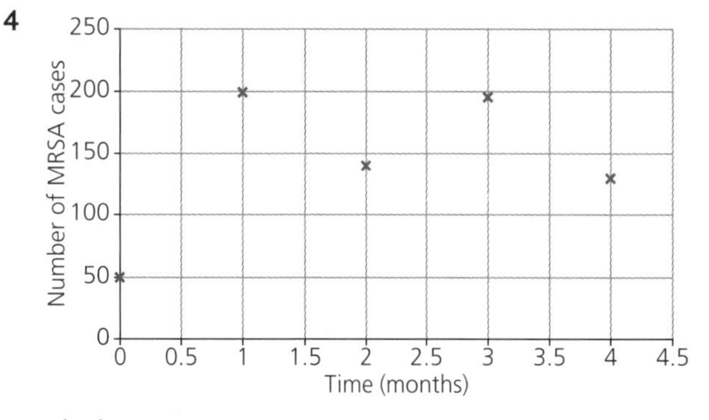

Graph showing MRSA cases

Ploting two variables from experimental or other data (pages 64–67)

Guided question

1 **Step 1** Draw and label the vertical axis with the letter y and horizontal axis with the letter x.

Step 2 For the y-axis, the grid is 12 cm high, so each 1 cm distance represents 1 unit.

For the x-axis, the grid is 12 cm, so each 1 cm distance represents 0.5 units.

Step 3 The first point is at the intersection where the vertical line at $x = 0$ meets the horizontal line at $y = 4.5$. The second point is at the intersection where the vertical line at $x = 1$ meets the horizontal line at $y = 6$.

Step 4 Repeat until all points are plotted.

Practice questions

2

3

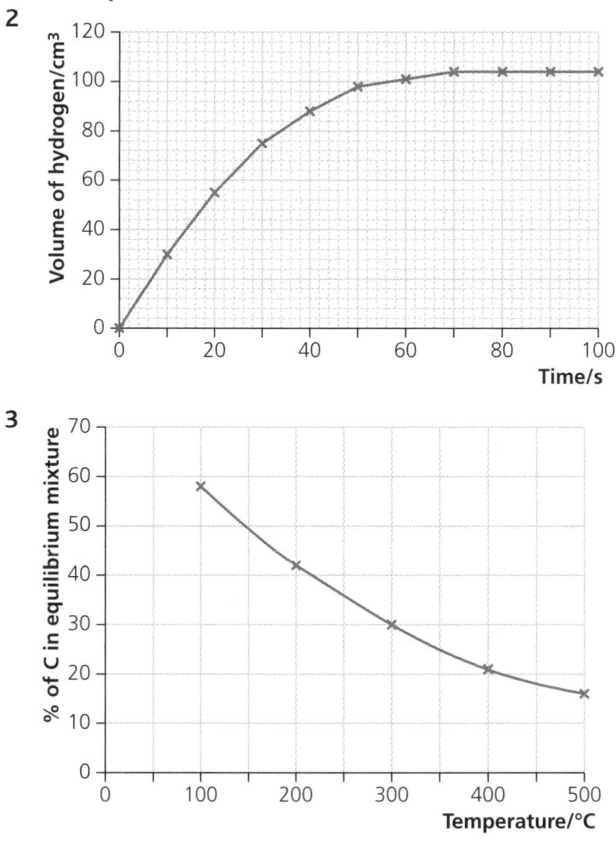

Determining the slope and intercept of a straight line (pages 67–71)

Guided question

1 a **Step 3** Graph A $\Delta y = 2 - (-4) = 6$

Graph B $\Delta y = 2 - (-2) = 4$

Graph C $\Delta y = 2 - (-2) = 4$

Step 4 Graph A $\Delta x = -3 - 1 = -4$

Graph B $\Delta x = -1 - 4 = 5$

Graph C $\Delta x = -2 - 1 = 3$

Step 5 Graph A: gradient $= -1.5$

Graph B: gradient $= +0.8$

Graph C: gradient $= -1.3$

b Graph A: intercept $= -2.5$; equation: $y = -1.5x \times -2.5$

Graph B: intercept $= -1$; equation: $y = 0.8x - 1$

Graph C: intercept $= -0.5$; equation: $y = -1.3x - 0.5$

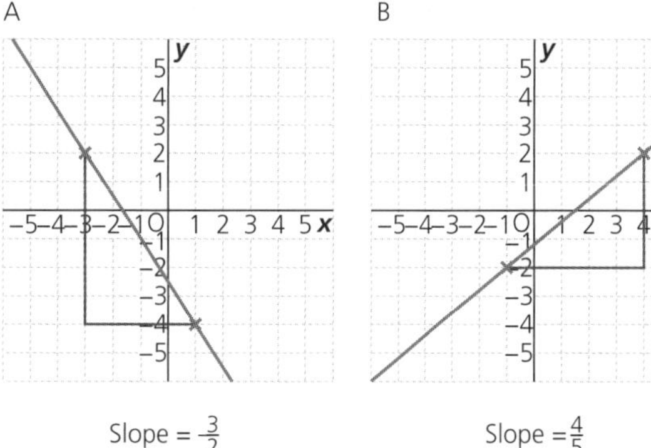

Slope $= -\frac{3}{2}$

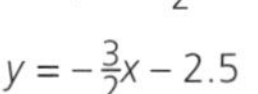

$y = -\frac{3}{2}x - 2.5$

Slope $= \frac{4}{5}$

$y = \frac{4}{5}x - 1$

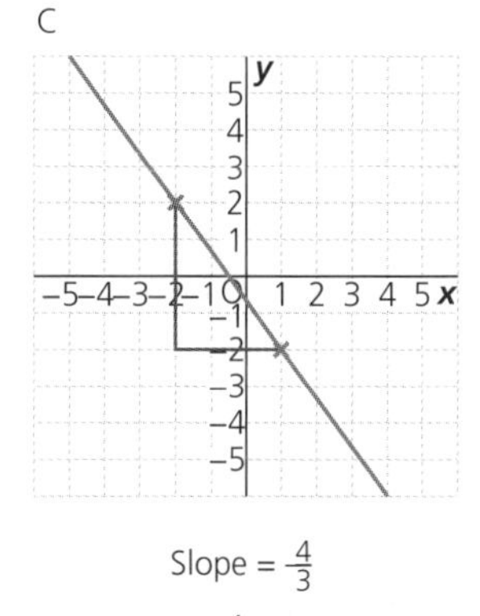

Slope $= -\frac{4}{3}$

$y = -\frac{4}{3}x - 0.5$

Practice questions

2 The internal concentration of the onion can be found by determining the point where there was no change in mass, as at this point the concentration inside the onion cell is equal to that of the external solution.

0% change in mass = 0.53 M

Therefore the internal concentration of the onion cell = 0.53 M

3 The rate of reaction is fastest at the start of the reaction. To find the rate of reaction, find the gradient of the line.

First, find the gradient of the line between 0 and 30 seconds:

Gradient = change in y ÷ change in x = 5 ÷ 30

Gradient = 0.17

Rate of reaction = 0.17 g/s

Drawing and using the slope of a tangent to a curve as a measure of rate of change (pages 72–75)

Guided question

1 **Step 3** gradient (m) = $\frac{\text{change in } y\text{-axis}}{\text{change in } x\text{-axis}} = \frac{\Delta y}{\Delta x} = \frac{1}{2}$

Practice questions

2

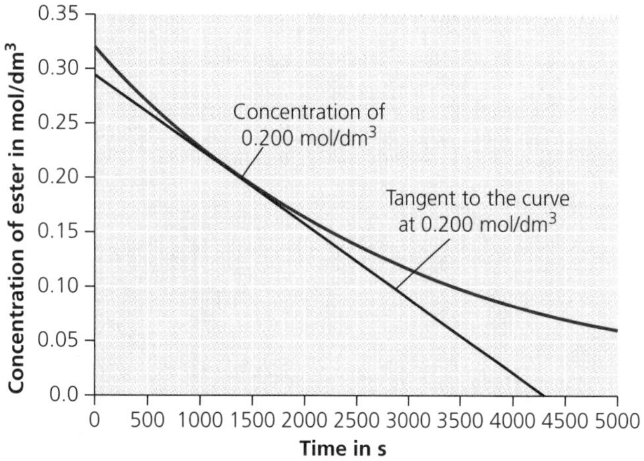

- At a concentration of 0.200 mol/dm³ the gradient of the tangent is $\frac{0.28}{3500} = 6.5 \times 10^{-5}$

3 a

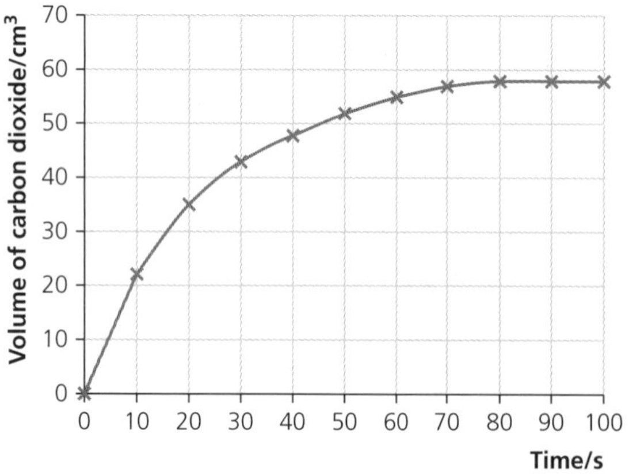

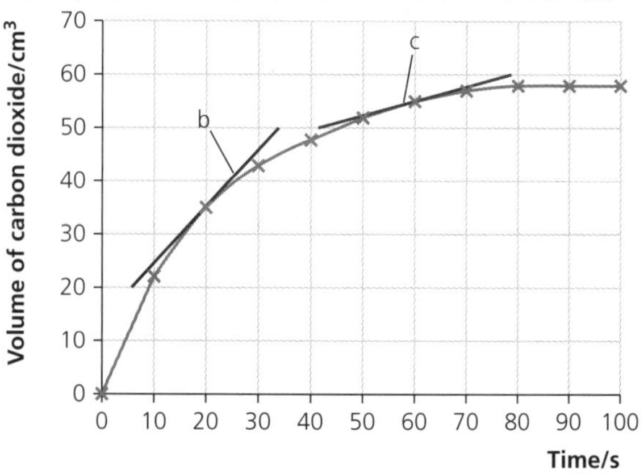

b gradient of tangent = $\frac{30}{28}$ = 1.1 = rate

c gradient of tangent = $\frac{10}{34}$ = 0.3 = rate

Geometry and trigonometry

Using angular measures in degrees (pages 76–78)

Guided question

1

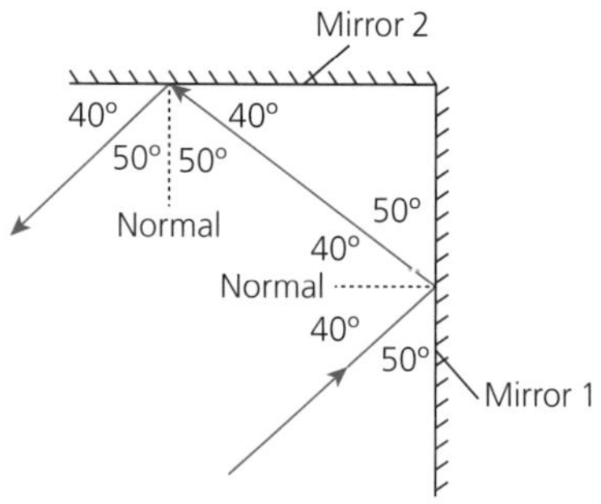

- **Step 4** Angle of reflection at Mirror 2: 50°

Practice question

2

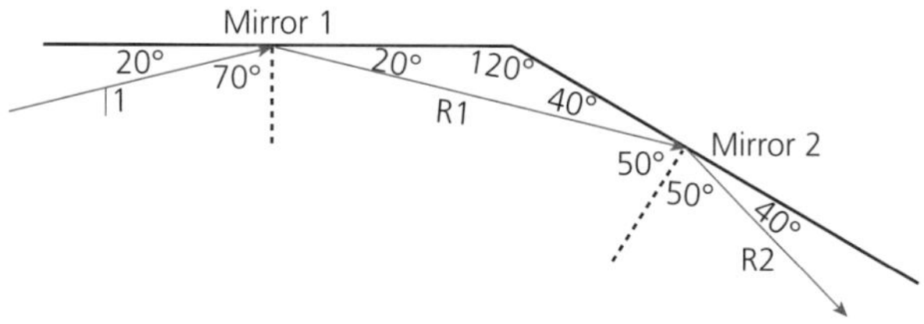

Since angles of incidence and refraction at Mirror 1 are both 70°, the glancing angle at the mirror is 20°.

Since angles in a triangle add up to 180°, the glancing angle at Mirror 2 is 40° and the angle of reflection at Mirror 2 is 50°.

Representing 2D and 3D forms (pages 78–81)

Guided question

1 H—N—H
      |
      H

Practice questions

2

3

O
H H

4 a

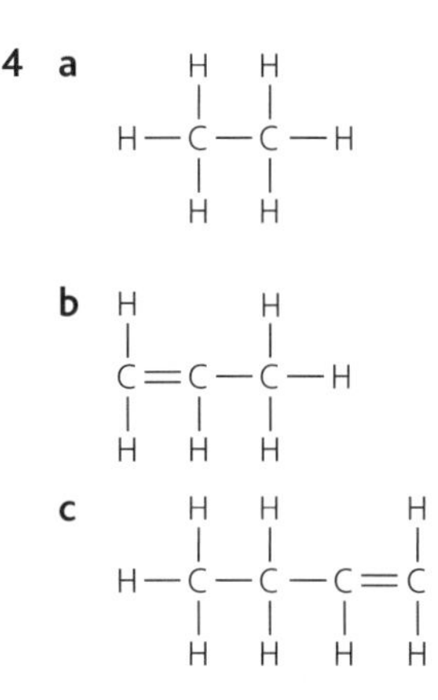

b

c

Calculating areas and volumes (pages 81–83)

Guided question

1 **Step 1** Surface area of cube = 8 × 8 × 6 = 384 mm^2

Step 2 Volume = 8 × 8 × 8 = 512 mm^3

Step 3 Surface area : volume = 384 : 512 = (384 ÷ 128) : (512 ÷ 128) as 128 is the largest common factor

Surface area : volume = 3 : 4

Practice questions

2 **Cube 1**

Surface area = side length × side length × number of sides
= 6 × 6 × 6

Surface area = 216 mm^2

Volume = length × width × height

Volume = 6 × 6 × 6 = 216 mm^3

Surface area : volume = 216 : 216
= (216 ÷ 216) : (216 ÷ 216)

Surface area : volume = 1:1

Cube 2

Surface area = side length × side length × number of sides
= 4 × 4 × 6

Surface area = 96 cm^2

Volume = length × width × height

Volume = 4 × 4 × 4 = 64 cm^3

Surface area : volume = 96 : 64
= (96 ÷ 32) : (64 ÷ 32)

Surface area : volume = 3 : 2

As 3 : 2 is a larger surface area : volume ratio than 1 : 1, cube 2 has the largest surface area : volume ratio.

3 surface area = 2 × 2 × 6 = 24 cm^2

volume = 2 × 2 × 2 = 8 cm^3

surface area : volume
24 : 8
3 : 1

surface area = 20 × 20 × 6 = 2400 cm^2

volume = 20 × 20 × 20 = 8000 cm^3

Surface area : volume
2400 : 8000 (divide by 3)
300 : 1000
3 : 10
0.3 : 1

The cube with small sides has a surface area to volume ratio that is ten times bigger.

» 2 Extended response questions

Extended responses: Describe (pages 89–90)

Expert commentary

1 This is a model answer that would get the full marks.

Connect a variable resistor, an ammeter, a power supply unit (PSU) and a 50 cm length of resistance wire in series with each other. Connect a voltmeter across the resistance wire.

Switch on the PSU and record the readings of current and voltage in a table. Switch off the PSU to allow the wire to cool down. Calculate the resistance by dividing the voltage by the current.

Switch on the PSU again, adjust the variable resistor and record the new current and voltage readings.

Switch off the PSU and calculate the wire's resistance. Then find the average resistance of the two calculations.

Repeat for other wires of increasing length.

Plot a graph of average resistance against length and draw the straight line of best fit. It should pass through (0,0), confirming that the resistance of the wire is directly proportional to its length.

This description is full of accurate detail and there is clear evidence that the candidate knows exactly what has to be done. A table showing how the results are recorded would have been helpful, but it is not essential.

The student shows knowledge and understanding as to how the apparatus is arranged, the readings that have to be taken and the precaution necessary to obtain satisfactory results. The candidate also knows how to find the resistance and how the result s must be processed in order to draw a conclusion.

Peer assessment

2 This answer would get a level of 1 and a mark of 2.

This is because the answer does include some correct points, such as the need for sperm and eggs to mix for fertilisation to take place and the use of FSH. However, it also contains a large number of errors. Two key mistakes are that the embryos are implanted into the mother's womb and not grown in a test tube, and also that FSH is a hormone, not an enzyme.

The answer is also not very well structured, with the use of FSH – which occurs at the start of the process – not mentioned until the end of the answer. There is also a grammatical error in the phrase 'FSH which are an enzyme' – it should be 'FSH which is an enzyme'.

Improve the answer

3 A model answer that would be awarded the full marks is:

I would place some zinc chloride in an evaporating basin. I would put two electrodes in the zinc chloride and attach one to the positive (the anode) of a power pack or battery and the other to the negative (the cathode). I would place the evaporating basin on a gauze and heat gently with a Bunsen burner. When the zinc chloride is beginning to melt, I would make sure the electrodes do not touch, and then switch on the power pack. At the positive anode a grey liquid is observed and at the cathode a green-yellow gas. A fume cupboard should be used.

Extended responses: Explain (pages 91–92)

Expert commentary

1 This is a model answer that would get full marks.

Copper is a metal and has lots of delocalised electrons that can move and carry charge and so it conducts electricity. Copper chloride is an ionic compound. It cannot conduct when it is solid because the ions are held tightly in place but when it is dissolved in solution then the ions can move and carry the charge. Chlorine is a molecule and it does not have charge and so it cannot conduct electricity.

Peer assessment

2 This answer would get a level of 3 and a mark of 5.

This is because it is a clear, well-structured answer that covers all the main points. There is, however, a key mistake, which is that lipids are not broken down into amino acids, but are actually broken into fatty acids and glycerol. This means that this answer scores the lowest mark in level 3. It shows the importance of ensuring that the key information in your answer is correct – even incorrectly explaining one key point can lose you marks.

Improve the answer

3 This is an improved answer that would get all 6 marks.

Pressure is the force acting on a surface divided by the area of the surface.

The column of liquid is a prism of cross-section area A and height h.

The volume of liquid is the product $A \times h$.

The mass of the liquid is $A \times h \times \rho$, *where* ρ *is the liquid density.*

The weight of the liquid is $A \times h \times \rho \times g$, *where g is the gravitational field strength.*

The pressure, P, is the weight of liquid divided by the area, so $P = h \times \rho \times g$.

Extended responses: Design, Plan or Outline (pages 92–95)

Expert commentary

1 This is a model answer that would get the full marks.

To test this hypothesis, first you would add a known concentration of penicillin to one disc, and a known concentration of tigecycline to another disc. Each disc containing antibiotic should have the same diameter. You would then place each disc in the centre of an agar plate to which a known volume of a specific concentration of bacterial culture had been added. You would then incubate both discs at 37°C for 24 hours.

In this time, the antibiotic will diffuse out of the disc and into the agar. A clear area will be produced where the bacteria is killed by the antibiotic. You should measure the clear area produced by each disc. Then you should repeat the whole investigation at least three times to calculate a mean clear area.

You should compare these two mean clear areas, and if the tigecycline produces a larger mean clear area than the penicillin, then this proves the hypothesis. If not, this disproves the hypothesis.

This is an excellent answer that will score all 6 marks available. The student identifies the independent variable correctly (the type of antibiotic) and states how to vary it (change the type of antibiotic on the disc).

Throughout the answer, control variables are identified correctly and controlled, for example, same diameter of disc, same volume and concentration of bacterial culture, time and temperature of incubation. The reliability of the investigation is also covered by suggestions to repeat the investigation and calculate a mean clear area.

The student ends well by relating back to the question and saying how the hypothesis could be proved or disproved.

Peer assessment

2 This question would be awarded a level 1 and a mark of 1. This is because credit can be awarded for indicative content points 3 and 5 only.

There are also several spelling/grammar mistakes. The candidate has used 'were' instead of 'where', 'mesure' instead of 'measure', 'angel' instead of 'angle' and there are no full stops. The quality of the written English is so poor that the candidate would probably be placed at the bottom of the mark band.

The student does not mention paper, protractor or ray box and therefore gains no credit for indicative points 1, 2 and 4. The angle of refraction is also wrongly identified and the marks for points 6 and 7 are therefore lost.

Improve the answer

3 This is an improved answer that would get all 6 marks.

First place some potassium iodide solution in a test tube and add some aqueous chlorine. If there is a colour change to a yellow/brown solution then iodine has been produced because chlorine is more reactive and will displace iodine from the solution.

In a second test tube, place some potassium iodide solution and add some aqueous bromine. If the solution turns to yellow/brown then iodine has been produced because bromine is more reactive and will displace iodine from the solution.

In a third test tube, place some potassium bromide solution and add some aqueous chlorine. If the solution turns orange/red-brown then bromine has been produced because chlorine is more reactive than bromine and will displace it from the solution.

Extended responses: Justify (pages 95–97)

Expert commentary

1 This is a model answer that would get the full marks.

Measles is a very serious disease that can be fatal. It is therefore very important to ensure that as many children as possible are vaccinated against measles, because this means they will not contract the disease themselves. By vaccinating a large proportion of young children, the vaccination also prevents measles from spreading.

Measles is a viral infection, so it cannot be treated with antibiotics. A more appropriate treatment is ensuring that the patient has enough fluids, and is resting. Measles is spread by inhaling droplets from sneezes and coughs, so if infected people are kept away from others, the infection is less likely to spread.

This is an excellent answer that would score all 6 marks available. All of the strategies listed in the table are fully justified using the student's scientific knowledge. This includes the importance of vaccinating a large proportion of young people to prevent measles spreading and the reason why it cannot be treated with antibiotics. The answer links keeping people away from public areas to how measles is spread.

The answer is also very well-structured, with each of these points dealt with in the order they appear in the table in the question.

Peer assessment

2 This answer would be awarded a level of 2 and a mark of 4. This is because the student only gains credit for indicative content points 1, 3, 4 and 5. There are also no spelling or grammar mistakes.

The student refers to what happens at 100 °C and 0 °C when the material changes state. This is irrelevant information because the question specifically referred to liquid water over the range 0 °C to 100 °C.

It is also partly wrong; ice is less dense than liquid water (that's why icebergs float). The student's attention should have been drawn to the minimum on the graph indicating maximum density at 4 °C.

Improve the answer

3 This is a model answer that would score all 6 marks.

The graph shows that increasing light intensity increases the rate of photosynthesis. By increasing the light intensity in the greenhouse, the farmer will increase the amount of photosynthesis their crops are doing. This means they will have a greater growth rate and the farmer will increase her yield.

At high light intensities, the graph levels out. This is because another factor is limiting the rate of photosynthesis and increasing light intensity no longer has an effect. Temperature is an example of another limiting factor of photosynthesis. By also increasing the temperature, the rate of photosynthesis will increase to an even higher rate than if light intensity alone had been increased. This means that the farmer's actions are all scientifically justifiable.

Extended responses: Evaluate (pages 98–99)

Expert commentary

1 This is a model answer that would get full marks.

The raw material to make hydrogen is water and there is an abundant supply, for example in seas and lakes. Diesel comes from crude oil, which is in limited supply and is running out. To save crude oil, a non-renewable resource, it is better to use hydrogen. Hydrogen is expensive to produce from water as electricity is needed and the generation of electricity may produce carbon dioxide, which contributes to the greenhouse effect, unless renewable power is used.

When hydrogen burns it produces water only and so it does not cause any air pollution, but diesel burns to produce carbon dioxide, which can increase the greenhouse effect. This causes global warming and ice caps to melt. Incomplete combustion of diesel may produce carbon monoxide, which is toxic, and also carbon, which can cause smog that causes respiratory problems. In conclusion, hydrogen is better to use because it is in good supply and does not cause pollution, but it is a flammable gas that is expensive to store safely.

Peer assessment

2 This would be awarded a Level 2 and given 3 marks.

This is Level 2 because an attempt has been made to describe some conditions, which comes to a conclusion for temperature. However, the logic may be inconsistent at times, particularly in terms of pressure, but it does build towards a coherent argument for temperature conditions.

The inaccuracies are:

- The student stated that increasing pressure did not have much effect; however, increasing pressure increases the yield.
- The student did not comment on the pressure to be used.

Improve the answer

3 This is a model answer that would score all 6 marks.

The addition of nitrate fertiliser would be a useful way to treat the chlorosis of the plants. Chlorosis is a condition partly due to a lack of proteins, and the plants could use the nitrates in the fertiliser to produce proteins. The nitrates would be taken into the roots by active transport and then transported in the xylem. The nitrates would be used to produce amino acids, which can then be used to synthesise proteins at the ribosomes inside the plant's cells.

Nitrate fertiliser would not treat the condition fully as chlorosis is also due to a lack of chlorophyll. Magnesium ions are required to produce chlorophyll so these ions would also need to be present in the fertiliser.

Extended responses: Use (pages 100–102)

Expert commentary

1 This is a model answer that would get full marks.

The problem is that some people in the audience hear the same sound several times with a very short time between each. The first people hear the sound directly from the performer and shortly afterwards they hear echoes. Since the echoes travel different distances, they will arrive at different times.

Echoes are reflections of sound waves, and come from hard surfaces on the walls and ceiling. Correcting the problem requires elimination of the hard, echo-producing surfaces. This can be achieved by fitting soft and pleated curtains to the walls. Soft surfaces are excellent sound absorbers and pleats give a very large surface area, which assists this absorption.

The ceilings might also be tiled with tiles containing a soft mineral wool, which is an excellent sound absorber.

This is a good answer because the information given in the text of the question and in the diagram is used to identify the problem, its cause and its solution.

Peer assessment

2 This would be awarded a Level 3 and given 5 marks.

This was a Level 3 answer because a detailed and coherent explanation is given that demonstrates a good knowledge and understanding and it refers to energy changes and temperature changes for all three substances, correctly deducing that a negative sign means heat is given out.

The inaccuracies include:

- The student writes that the temperature got colder. It would be more accurate to write that the temperature decreases.
- The last sentence about time is irrelevant.

Improve the answer

3 This is an improved answer that would get all 6 marks.

Ionisation occurs when an atom or molecule becomes charged. This occurs when it gains or loses electrons. In this case the atom or molecule is much more likely to become positively charged as the colliding alpha particle will knock out some of the orbiting electrons.

Smoke in the chamber displaces the air between the alpha source and the detector. This reduces the amount of ionisation that occurs and fewer ions reach the detector, causing current to be sent to the alarm circuit and the alarm to ring.

Beta particles and gamma rays have much lower ionising ability than alpha particles in air and neither would work in this type of smoke detector. Americium-241 is the best choice because it will never have to be replaced during the lifetime of the detector.

» 3 Working scientifically

The development of scientific thinking (pages 104–109)

1 The scientific method is the formulation, testing and modification of hypotheses by systematic observation, measurement and experimentation.

2 They carry out experiments.

3 Charles Darwin used his own observations, experimentation and the developing knowledge of geology and fossils to develop his theory of evolution by natural selection. This differed from other, older theories such as Lamarck's, which stated that changes that occur during an organism's lifetime can be inherited. New evidence such as understanding the mechanisms

of inheritance have led to the theory becoming widely accepted.

4 Models are important to explain and describe phenomena in an understandable way, as well as to make predictions.

5 Some people consider an embryo to be a potential life and therefore to have a right to life.

6 The impact of overfishing can be reduced through technology by using nets with large mesh sizes to allow small, young fish to escape.

7 Burns/scalds from hot water and steam.

8 Ethanol is flammable so do not heat it directly; instead use a hot water bath.

9 **a** Carrying out an experiment; drawing conclusions/reporting results.

b

Risk	Control measure
Potassium nitrate powder is an irritant	Wear gloves/wash hands immediately if some falls on skin

10 The results should be peer assessed. This means they should be evaluated by other scientists working in the same field.

Experimental skills and strategies (pages 109–118)

1	Independent	Dependent	Controlled
a	length	resistance	cross-sectional area (or material from which wire is made)
b	force	acceleration	mass of trolley
c	mass of copper carbonate	time taken to react and disappear	surface area of copper carbonate; volume and concentration of acid; temperature
d	mass of calcium carbonate	volume of carbon dioxide	surface area of calcium carbonate; volume and conc. of HCl; temperature
e	distance of light from pondweed	number of bubbles produced in five minutes	species of pondweed, mass of pondweed

2 **a** speed of dissolving depends on temperature

b blue solution formed

c beaker/test tube; stirring rod; stopwatch

3 The student should add missing part of glassware attached to bung to the diagram – copper(II) sulfate solution in (conical) flask/boiling tube attached to glassware with the bung

Pure water in test tube/flask/beaker at end of delivery tube, which must not be sealed

Heat source to heat container holding copper(II) sulfate solution

4 **a** top-pan balance

b burette

c stopwatch (digital or analogue)

d milliammeter (digital or analogue)

5 Errors in methodology are due to a mistake in the planning of an experiment leading to results not being accurate or precise. Errors in carrying out the investigation are those occurring as the plan is being carried out, not due to the plan being incorrect.

6 Taking a number of different samples increases the chance of the results being representative.

7 **a** magnesium + copper sulfate → magnesium sulfate + copper

b The student has stated what has happened in the experiment but has not given observations. The correct observations are red brown solid forms or the blue solution changes in colour to colourless.

8 **a** A suitable format might be.

Volume (cm^3)	20	35	45	50	55
Mass (g)	16	28	36	40	42

b The student might want to repeat the 55 cm^3 and 42 g results, because all other pairs of values give a density of 0.80 g/cm^3, but this pair of values gives a density of 0.76 g/cm^3.

9

Time (s)	10	20	30	40	50
Volume 1 (cm^3)	30	49	59	63	63
Volume 2 (cm^3)	32	51	59	63	65
Average volume (cm^3)	31	50	59	63	64

10 **a** Stopclock

b **i** Student C

ii Student A

c It will improve reliability.

Analysis and evaluation (pages 118–120)

1 Anomalous results are values that are very different to the rest of the results of an investigation.

2 a Student A = 21.3 ± 0.2%, Student B = 22.5 ± 0.1%

b Student A was accurate, Student B was not accurate

c Both students had repeatable results

d Due to random errors

e Student B had a systematic error; results were consistently about 1.3% too high

3 a The result at 4.5 minutes

b 100.3 g

c As time increases the mass of the flask and contents decreases. The mass decreases more rapidly from 103.0 to 99.4 g between 0 and 4 minutes then it gradually decreases until at 7 minutes it is constant at 99.0 g.

d Repeat more times and calculate the mean

e Use a different balance

4 a Systematic

b Random

c Systematic

5 An error is caused by a defect in the apparatus, technique or inconsistency in the measurement. A mistake is caused by a person using apparatus (for example, a calculator) incorrectly.

6 It reduces random error where some results are too big, and others too small. Repeating and averaging cancels out the small values with the big values.

7 a All the results are smaller than the true value, so this is a systematic error.

b Reproducible and valid

8 Repeatable results are carried out under the same conditions and by the same investigator so the same systematic error may occur each time. As reproducible results are gathered by different investigators with different equipment, it is less likely they will make the same systematic errors.

9 Large range bars indicate uncertain results as the range of results around the mean is large.

» Exam-style questions

Biology Paper 1 (pages 150–153)

1 a Gametes [1]

b The number of chromosomes halves during meiosis [1] so the daughter cell has half the number of chromosomes compared with the parent cell after the point meiosis has occurred [1].

c By changing to have half the number of chromosomes, this means that when the two gametes fuse during fertilisation [1], the cell formed has the full and correct number of chromosomes [1].

d Award [1] mark for initial correct number of chromosomes, [1] mark for maintaining the number of chromosomes after mitosis.

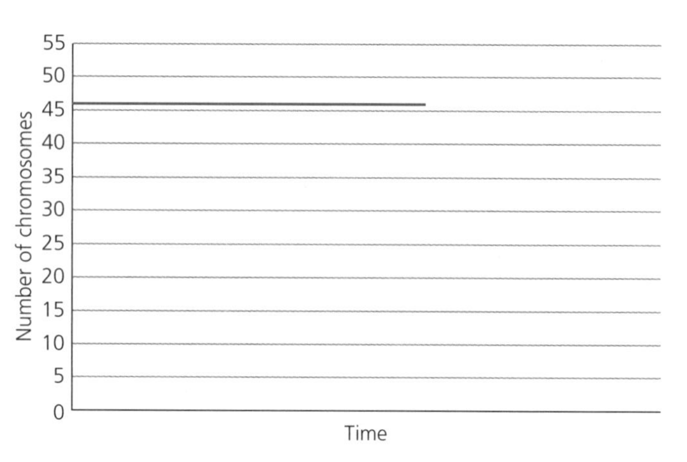

Graph to show the chromosome number in a cell before and after mitosis has occurred

2 a 1500 μm = 1.5×10^2 [1 mark for correct significant figures, 1 mark for correct use of standard form.]

b The artery has a thick wall to withstand the high pressure of blood flow [1]. Veins carry blood at lower pressure, so have thinner walls [1]. Capillaries have very thin walls to allow diffusion of gases and other substances into and out of them [1].

c Arteries are affected by coronary heart disease [1]. As the coronary arteries become blocked, the heart does not receive oxygen [1].

3 a Surface area : volume = 100 : 0.002 [1]

= (100 × 500) : (0.002 × 500) = 50 000 : 1 [1]

b Surface area : volume = 1.8 : 0.1 [1]

= (1.8 ×10) : (0.1 ×10) = 18 : 1 [1]

c The surface area : volume ratio for the alveoli is much larger than that of the human [1]. This shows that diffusion through the surface would not be fast enough [1] to meet the needs of the human, therefore they require alveoli [1].

d 15 mm = 15 000 μm [1]

Magnification = size of image ÷ size of object

Magnification = 15 000 ÷ 2 [1]

Magnification = 7500 [1]

e No [1]. Surface area and thickness of diffusion membrane are given in the question, and these are required to calculate Fick's law [1]. However, concentration difference is also required, and this is not given [1].

f The lungs have an efficient blood supply [1] and are also ventilated [1].

4 a DNA ligase joins [1] two sections of DNA from different organisms together [1].

b Restriction enzymes cut DNA [1] at specific points [1].

5 Award marks based on the indicative content provided, up to a maximum of 6 marks:

Indicative content:

- Recruit participants who are same gender, age, fitness level.
- Measure heart rate before exercise to determine resting level.
- Ensure heart rate is at resting level before recording heart rate.
- Participants should exercise in the same way (for example running on the spot) for a set period of time (for example 2 minutes).
- Measure time taken to return to resting heart rate for each participant.
- Use times to calculate a mean time to return to resting heart rate.
- After a set period of rest, repeat the investigation with longer exercise periods (for example, 4 minutes, 6 minutes, 8 minutes and 10 minutes).
- Compare the mean times taken to return to resting heart rate for the different periods of exercise.

6 a Award marks as follows:

- [1] for a suitable scale on the axis.
- [1] for correctly labelled axis.
- [2] for 5 points correctly plotted.
- [–1] for each error.

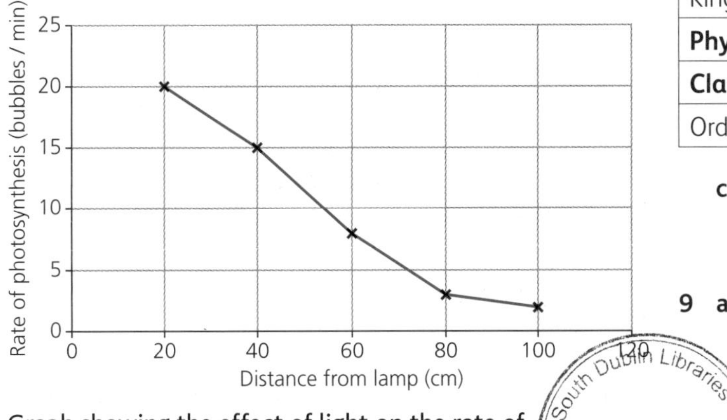

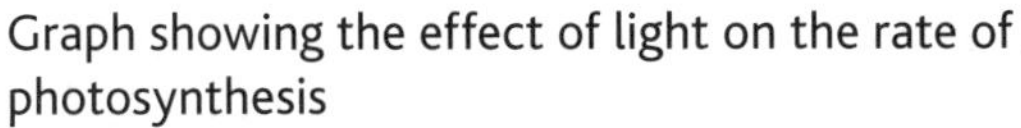
Graph showing the effect of light on the rate of photosynthesis

b i [1]

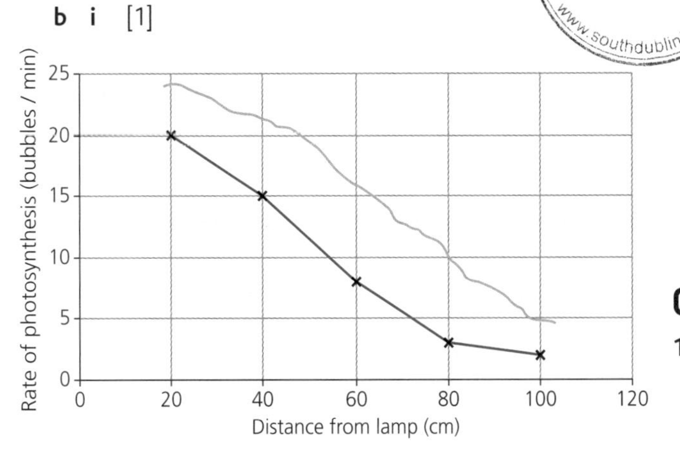

Graph showing the effect of increasing the temperature on the results

ii The rate of photosynthesis would be higher at each distance as temperature is limiting the rate of reaction [1]. Increasing the temperature will lead to a higher rate of photosynthesis [1].

7 a Proteins are synthesised on ribosomes [1] so if the tRNA is blocked from transferring amino acids to the ribosome the amino acids will not be added to the growing protein chain [1]. This means the proteins the bacteria need for growth will not be produced [1].

b $r = \text{diameter} \div 2$

$r = 6 \div 2 = 3\,\text{cm}$

$A = \pi r^2$

$A = \pi \times 3^2$ [1]

$A = \pi \times 9 = 28.3\,\text{cm}^2$ [1]

c HIV is caused by a virus [1]. Viruses cannot be treated with antibiotics [1].

d Unlike linezolid, penicillin is a natural antibiotic [1], which was discovered rather than synthesised as it is produced by the *Penicillium* mould [1].

8 a The first word of the name is the genus of the mussel (*Mytilus*) [1].

b Award [1] mark for each correct answer in bold below.

Kingdom	**Animal**
Phylum	Mollusca
Class	Bivalvia
Order	Ostreoida

c The blue mussel is in the eukaryote domain [1] as it is an animal [1] OR it has genetic material enclosed in a nucleus [1].

9 a i The stonefly population will decrease [1], and the diatom and algae population will increase [1].

ii The population of leech and stonefly will decrease [1], which could lead to an increase in the shrimp and mayfly population [1].

b i producer – diatoms or algae [1]

ii primary consumer – shrimp or mayfly [1]

iii secondary consumer – leech or stonefly [1]

c An increase in sludge worms would indicate that the water is polluted [1] as sludge worms are an indicator species [1].

Chemistry Paper 1 (pages 153–157)

1 a 2 marks for plotting points correctly, 1 mark for a smooth line. [3]

b The temperature increases so heat was given out. [1]

c Burette [1]

d Use a lid on the cup with a hole to allow the burette in [1]; this prevents heat loss [1].

OR

Use a pipette to measure the acid [1]. It is accurate to one decimal place [1].

e $NaOH + HCl \rightarrow NaCl + H_2O$ [2]

f $25 \times \frac{0.10}{1000} = 0.0025\,\text{mol}$ [1]

2 a Bubbles (due to the carbon dioxide produced) [1]

b Examples may include: Wear safety glasses [1]; tie long hair back [1].

c M_r $MgCO_3 = 84$, M_r $MgSO_4 = 120$ [1]

$\frac{2.1}{84} = 0.025\,\text{mol}\ MgCO_3$ [1]

$0.025 \times 120 = 3.0\,\text{g}\ MgSO_4$ [1]

d $\frac{1.8}{3.0} \times 100$ [1] $= 60\%$ [1]

e Some may be lost in filtering or during transfer between apparatus. [1]

3 a phenolphthalein (colourless to pink)/methyl orange (yellow to red) [1]/litmus (blue to red) [1]

b Colour change would not be visible in red wine. [1]

c i $18.90 + 19.00/2$ [1] $= 18.95\,\text{cm}^3$ [1]

ii $18.95 \times \frac{0.100}{1000}$ [1] $= 0.001895\,\text{mol NaOH}$ [1]

2 mol NaOH : 1 mol tartaric acid

$\frac{0.001895}{2} = 0.0009475\,\text{mol tartaric acid}$ [1]

$0.000975 \times \frac{1000}{25.0}$ [1] $= 0.379$

$= 0.038\,\text{mol/dm}^3$ [1]

iii 150 [1]

4 Award 1 mark for each of the indicative content points covered below, up to a maximum of 6 marks.

Indicative content:

- thermometer
- measuring cylinder/pipette
- spatula
- plastic cup (with lid)
- weigh the same mass of each metal in same state of division, e.g. powder
- measure a volume of sulfuric acid into a plastic cup
- measure and record the temperature of the sulfuric acid
- add metal P into the plastic cup
- stir and record the highest temperature
- repeat for each metal at least three times to calculate a mean
- calculate the mean temperature change; the greatest temperature change shows the most reactive metal

5 a A: Gas syringe [1]; B: Conical flask [1]

b $Mg + 2HCl \rightarrow H_2 + MgCl$ [2]

c Advantage: convenient/quick to use [1]

Disadvantage: inaccurate/only accurate to $1\,\text{cm}^3$ [1]

d Award marks as follows: sensible scales, using at least half the grid for the points [1], all points correct [1], best-fit line [1]

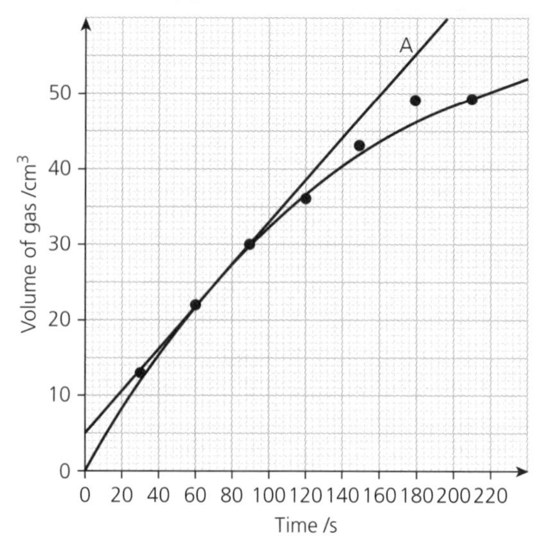

e 80 s (read from graph) [1]

f Volume of gas $= 13\,\text{cm}^3$ [1]

Mean rate $= \frac{13}{30} = 0.433\,\text{cm}^3/\text{s}$ [1]

$= 0.4$ [1] cm^3/s (to 1 s.f.) [1]

6 a Total mass $= 37.5\,\text{g}$ [1]

$\%\,Sn = \frac{15}{37.5} \times 100 = 40\%$ [1]

b C [1]

7 a So that it does not run on the paper [1]

b Three [1]

c Distance moved by the blue spot $= 3.3\,\text{cm}$ [1]; distance moved by solvent $= 4.4\,\text{cm}$ [1];

R_f value $= \frac{3.3}{4.4} = 0.75$ [1]; Blue spot = D [1]

8 a $6CO_2 + 6H_2O \rightarrow C_6H_{12}O_6 + 6O_2$; Award [1] mark for $C_6H_{12}O_6$ and [1] mark for balancing.

b Carbon dioxide dissolves in seawater and is used by sea life to form calcium carbonate/shells [1], which becomes sediment and forms sedimentary rock [1].

c i It has increased [1]

ii $\frac{(390 - 370)}{370} \times 100$ [1] $= \frac{20}{370} \times 100 = 5.41\%$ [1]

iii increased burning of fossil fuels [1], deforestation [1]

9 Award 1 mark for each of the points below, up to a maximum of 4 marks.

- In Earth's atmosphere there is much less carbon dioxide – 0.04%. [1]
- In Earth's atmosphere there is much more nitrogen – 78%. [1]
- In Earth's atmosphere there is much more oxygen – 21%. [1]

- In Earth's atmosphere there is no methane. [1]
- Both have noble gases in small amounts. [1]

Physics Paper 1 (pages 157–160)

1 a area = length × breadth

$= 0.5\,m \times 0.4\,m$ [1]

$= 0.2\,m^2$ [1]

b volume $= 0.01\,cm^3 = 0.01 \div 1\,000\,000$

$= 1 \times 10^{-8}\,m^3$ [1]

c diameter = volume ÷ area

$= 1 \times 10^{-8} \div 0.2$ [1] (allow error carried forward)

$= 5 \times 10^{-8}\,m$ [1]

d The oil on the surface of the water cannot be *less* than 1 molecule thick, suggesting that the measured diameter is an overestimate [1].

2 a i thickness = 47 mm ÷ 500 [1] = 0.094 mm [1]

ii By measuring to the nearest mm, the thickness of the 500 sheet ream t is known to be $46.5 \leqslant t < 47.5\,mm$. [1]

The minimum thickness of a single sheet = (46.5 mm) ÷ 500 = 0.093 mm (to 2 d.p.) [1]

b i Wrap about 20 turns of wire on a pencil. Push the turns of wire together to form a tight coil [1]. Measure the length of the coil with a ruler [1]. Divide the length by the number of turns to find the wire's thickness [1].

ii Mass of wire [1] and length of wire [1].

3 a i Energy content = 40 litres × 32 MJ per litre [1] = 1280 MJ [1]

ii Energy converted to heat = (7 ÷ 10) × 1280 MJ = 896 MJ [1] (allow error carried forward).

iii Useful energy = (9 ÷ 10) × (1280 – 896) = 345.6 MJ [1] (allow error carried forward).

iv Efficiency = useful output energy ÷ total input energy = 345.6 ÷ 1280 [1] = 0.27 [1] (allow error carried forward).

b i 20 litres petrol contains 0.5 × 1280 MJ = 640 MJ

Additional energy in petrol = (640 – 150) MJ [1] = 490 MJ [1] (allow error carried forward).

ii We do not know the mass of the car or its load carrying capacity. [1]

iii Award 1 mark for each of the indicative content points covered below, up to a maximum of 6 marks. Accept any other reasonable answers.

Disadvantages

- Batteries have a much lower energy density than petrol (they store fewer joules per kilogram), so battery cars have a much smaller range than petrol cars.
- Batteries for electric cars are still being developed, so electric cars are more expensive than petrol cars as manufacturers try to recover the development costs.
- It is said that battery cars are less polluting than petrol cars, but this takes no account of the additional pollution that might be created by producing the electricity to drive them. (It is thought that if all the cars in the UK were electric cars, the UK would require the equivalent of 10 additional nuclear power stations for these cars alone.)
- Battery cars use rare earth metals from the Earth's crust and there is currently no way known to recycle them.

Advantages

- Electric cars produce no CO_2, so no greenhouse gases are produced by the cars themselves.
- No oxides of nitrogen or particulate pollutants are produced by electric cars – these pollutants cause serious health problems.
- The engine of an electric car is much more efficient than that of a petrol car, so less of the Earth's resources are consumed by the cars themselves.

4 a $P = (600 - 378) \div 600$ [1] = 0.37 [1]

b Both axes labelled with units as in the table [1]; scales chosen to cover at least half of each axis [1]; scales chosen allow for a straightforward interpolation [1]; points plotted to within 1 small square [2] (deduct $\frac{1}{2}$ mark for each incorrect or missing point – round up mark if necessary); smooth curve drawn through the data points [1].

c See graph for working. [1]

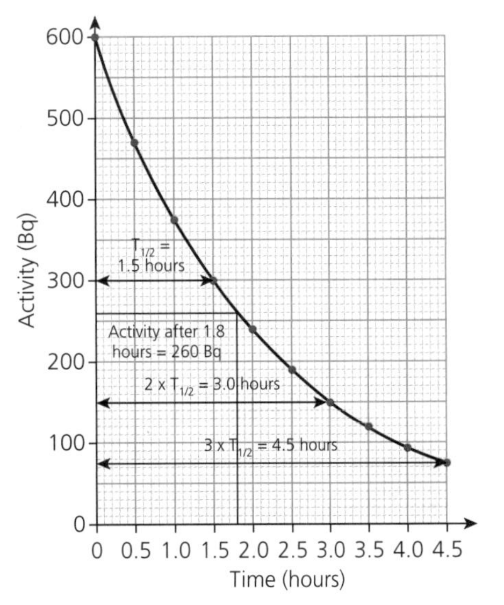

d Evidence from graph – vertical line at 1.8 hours to curve, horizontal line to vertical axis. [1]

Activity = 260 Bq [1]

5 a Award one mark for each correctly calculated current.

Resistance / Ω	5	7	8	4
Current / mA	480	240	160	320

b There are three parts to the network:

1. a single resistor of 5 Ω
2. two 7 Ω resistors, with combined resistance of 3.5 Ω [1]
3. an 8 Ω resistor and a 4 Ω resistor, with combined resistance of 2.67 Ω [1]

The voltage is greatest across the combination which has greatest resistance: the single 5 Ω resistor [1].

c Total current in network doubles [1], so current in 8 Ω resistor also doubles [1].

6 a i B is the primary coil [1]. The transformer reduces the voltage from 12 kV to 24 V and is, therefore, a step-down transformer [1]. Step-down transformers have more turns on the primary coil than the secondary coil [1].

ii turns ratio = voltage ratio = V_s:V_p
= 24:12 000 = 1:500 [1]

$$N_s = N_p \times \frac{V_s}{V_p} = 25000 \times \frac{1}{500}$$ [1]

= 50 turns [1]

b i Generator (in power station) [1]

ii A [1]

iii In Snowdonia, the electricity is transmitted underground [1] because the pylons are considered to be unsightly and would spoil the magnificent scenery [1]. This is not usually done because underground transmission is so expensive [1].

7 a i $P = \rho \times g \times h$ [1]

$7.35 \times 10^6 = 1050 \times 10 \times h$ [1]

$h = 700$ m [1]

ii The total pressure also includes the pressure of the air above the water. [1]

b i Volume increases because the pressure decreases. [1]

ii By Boyle's Law: $p_1V_1 = p_2V_2$ [1]

So, $(6 \times 10^6 \times 0.1) = (1 \times 10^5 \times V_2)$ [1]

which gives:

$$V_2 = \frac{6 \times 10^5}{1 \times 10^5}$$

$= 6\,\text{cm}^3$ [1]

iii An increase in water temperature would lead to an increase in the bubble volume in accordance with the gas laws. [1]

Key terms

Accuracy: How close we are to the true value of a measurement.

Active revision: Revision where you organise and use the material you are revising. This is in contrast to passive revision, which involves activities such as reading or copying notes where you are not engaging in active thought.

Angle of incidence, *i*: Angle between the incident ray and the normal to the boundary of a transparent material.

Angle of reflection: Angle between a reflected ray and the normal.

Angle of refraction, *r*: Angle between the refracted ray and the normal to the boundary of a transparent material.

Avogadro's number: The number of atoms, molecules or ions in one mole of a given substance.

Bar charts: Charts showing discrete data in which the height of the unconnected bars represents the frequency.

Base units: The units on which the SI system is based.

Categorical data: Data that can take one of a limited number of values (or categories). Categorical data is a type of discontinuous data.

Categoric variables: Variables which are not numeric (such as colour, shape).

Causal relationship: The reason why one quantity is increasing (or decreasing) is that the other quantity is also increasing (or decreasing).

Command word: An instructional term that tells you what the question is asking you to do. 'Describe' and 'Explain' are two examples of command words.

Common factor: A whole number that will divide into both the numerator and denominator of a fraction to give whole numbers.

Continuous data: Data that can have any value on a continuous scale, for example length in metres.

Continuous scale: A scale that has equal spaced increments.

Continuous variables: The variables which can have any numerical value (such as mass, length).

Control variables: Variables other than the independent variable that could affect the dependent variable, and are therefore kept constant and unchanged.

Critical angle: The angle of incidence in an optically dense medium when the angle of refraction in air is 90°.

Decider figure: The integer after the number of decimal places required, which decides whether we must round up or not.

Decimal places: The number of integers given after a decimal point.

Denominator: The number on the bottom of the fraction.

Dependent variable: The variable measured during an investigation.

Derived units: Combinations of base units such as m/s and kg/m^3.

Directly proportional: Quantities x and y are directly proportional to each other if their ratio $y:x$ is constant.

Discontinuous data: Data that can have a limited range of different values, for example eye colour.

Discrete data: Data that can only have particular values, such as the number of marbles in a jar.

Ecological: The relation of living organisms to one another and to their physical surroundings.

Ethics: This is the consideration of the moral right or wrong of an action.

Evaluate: This means to weigh up the good points and the bad points.
Extrapolate: Extending a graph to estimate values.
Fair test: A test in which there is one independent variable, one dependent variable and all other variables are controlled.
Fraction: A number that represents part of a whole.
Geometry: The branch of mathematics concerned with shapes and size.
Gradient: This is another word for 'slope'. It is the change in the y-value divided by the change in the x-value.
High order skill: A challenging skill that is difficult to master but has wide ranging benefits across subjects.
Histograms: Charts showing continuous data in which the area of the bar represents the frequency.
Holistic: When all parts of a subject are interconnected and best understood with reference to the subject as a whole.
Hypotenuse: The longest side of a right-angled triangle.
Hypothesis: A proposed explanation for a phenomenon used as a starting point for further testing.
Incident ray: A ray that strikes a surface.
Independent variable: The variable selected to be changed by an investigator.
Integers: These are whole numbers, which includes zeros.
Intercept: This is the point where the graph crosses an axis. In the equation: $y = mx + c$, the y-intercept is where the graph crosses the y-axis when $x = 0$; it is the value for y when $x = 0$.
Inversely proportional: Quantities x and y are inversely proportional to each other if their product xy is constant.
Leading zero: A zero before a non-zero digit, for example 0.6 has one leading zero.
Mean: The mean is a type of average. Means are covered on pages xx.
Multiples: Large numbers of base or derived units, such as kilo- in kilogram.
Negative correlation: This occurs if one quantity tends to decrease when the other quantity increases.
No correlation: There is no relationship whatever between two quantities.
Normal: A line drawn at right angles to a surface.
Numerator: The number on the top of the fraction.
Origin: The start of an axis of a graph.
Outlier: A data point that is much larger or smaller than the nearest other data point.
Parallax error: A difference in the apparent value or position of an object caused by different lines of sight.
Peer review: The process by which experts in the same area of study evaluate the findings of another scientist before the research is considered for inclusion in a scientific publication.
Place value: The value of a digit in a number, for example in 926, the digits have values of 900, 20 and 6 to give the number 926.
Positive correlation: This occurs if one quantity tends to increase when the other quantity increases.
Precision: Precise measurements are those where the range is small.
Qualitative: Descriptions of how something appears, rather than with figures or numbers.
Quantitative: Measurements such as mass, temperature and volume, involve a numerical value. For these quantitative measurements, it is essential that units are included because stating that the mass of a solid is 0.4 says very little about the actual mass of the solid – it could be 0.4 g or 0.4 kg.

Random error: An error that causes a measurement to differ from the true value by different amounts each time.
Recurring: When a number goes on forever.
Reflected ray: A ray that is reflected from a surface.
Refractive index: The ratio sin i : sin r.
Relative formula mass, M_r: The sum of the relative atomic masses (A_r) of all the atoms shown in the formula.
Reliability: Where different people repeat the same experiment and get the same results.
Representative data: Sample data that is typical of the overall area or population being sampled.
Resolution: The fineness to which an instrument can be read.
Scatter graph: A graph plotted between two quantities to see if there might be a relationship between them.
Scientific method: The formulation, testing and modification of hypotheses by systematic observation, measurement and experiment.
Spurious digits: Digits that make a calculated value appear more precise than the data used in the original calculation.
Submultiples: Fractions of a base unit or derived unit such as centi- in cm.
Systematic error: An error that causes a measurement to differ from the true value by the same amount each time.
Tangent: This is a straight line that just touches the curve at a given point and does not cross the curve.
Trailing zeros: Zeros at the end of a number.
Trigonometry: The branch of mathematics concerned with the lengths and angles in triangles.